Wings of Mystery

BOOKS BY DALE TITLER:

WINGS OF MYSTERY
WINGS OF ADVENTURE
THE DAY THE RED BARON DIED
I FLEW A CAMEL (with M. Curtis Kinney)

For young readers:
UNNATURAL RESOURCES
HAUNTED TREASURES

Acknowledgment is made to the following for permission to reprint material from their publications: G. P. Putnam's Sons for the use of material from *The Aces* by Frederick Oughton, Copyright © 1960 by Frederick Oughton; *The New York Times* for the use of material from the issue of April 23, 1918; *Argosy* Magazine for my article "The Plane Designed to Bomb New York," which first appeared in that magazine in 1962.

Copyright © 1962, 1966, 1981 by Dale M. Titler
All rights reserved
No part of this book may be reproduced in any form
without permission in writing from the publisher
Printed in the United States of America

1 2 3 4 5 6 7 8 9 10

Library of Congress Cataloging in Publication Data

Titler, Dale Milton, 1926-
  Wings of mystery.

  Includes index.
  1. Aeronautics—History. 2. Aeronautics—Flights.
I. Title.
TL515.T52 1981   629.13'09   81-810
ISBN 0-396-07826-5   AACR2

Wings of Mystery

TRUE STORIES OF AVIATION HISTORY

REVISED EDITION

Dale Titler

Illustrated with photographs and maps

DODD MEAD & COMPANY
NEW YORK

For a distinguished warbird and gentleman,
who gave me a glimpse into the Golden Era of Flight
Oliver Colin LeBoutillier

Preface

There is a part of aviation history in our twentieth century—which is to say, most of it—that has a grand assortment of mysteries. They persist, stubbornly unresolved, a monument to man's efforts to conquer the air ocean. This is a book about some of them, a selection of true flying stories that stirred imaginations in the past and still do today. Most of them defy solution. Those that were flights to oblivion will never be solved, for the sparse clues left behind are as nebulous as the atmosphere through which the fliers and their machines once passed.

The stories are representative. "The Legion of Lost Warbirds" tells of missing military pilots of all air wars; "The Dole Derby" and the ocean fliers recall those who failed to conquer the seas by long-distance flight. The *Hindenburg* enigma brings to mind the dirigible disasters of war and peace and, to remind us of the dead men and broken aircraft still lost on the earth's untrodden stretches, there are the *Southern Cloud,* the *Lady Be Good,* and Captain Lancaster's *Southern Cross Minor,* recently recovered from the Central Sahara.

But in a few cases—very few—there may be a flicker of hope for solution. Because of the widespread interest in Paul Redfern's flight, Glenn Miller's disappearance, and the fate of Amelia Earhart and Fred Noonan, the answers to these mysteries may one day be uncovered.

Some of the accounts will strike the reader as a bit uncommon—even for aerial mysteries. I can't explain the odd details; I haven't tried. Nor can I explain the curious supernatural twists that have somehow found a place in certain chapters. I don't believe anyone can.

Although man has walked on the moon and sent space probes to the outer reaches of our solar system and beyond, we still face a great many unanswered questions about our own world. When pilots fly alone up there they get to wondering about them. I know I have. At times I've questioned my right to be there. And often, in looking down on the miles of land and water slipping under my wings, I've wondered whether the long road to the stars will be as fraught with byways beyond our human knowledge and understanding.

These accounts are backed with facts, official records, and the narratives of those who have lived them. But unlike most flying stories, they have few conclusions or real endings. This book will, in fact, leave you hanging.

Here's to some interesting hours of speculation anyway. . . .

DALE TITLER

Acknowledgments

This book can be traced back to the summer of 1940 when, as a fourteen-year-old nurturing the drive to one day become a pilot, I began to collect unusual flying stories. Later, when I took to writing, my scrapbook gave me the ideas that eventually became *Wings of Mystery*—as first published in 1966. During the past 14 years new developments have come to light and this revision updates earlier accounts and replaces others.

Many thoughtful and patient persons helped me collect the materials that went into the book—and into this updated version as well. To my collaborators go my thanks.

Special acknowledgement is due Captain (Ret.) Frank McGuire, late of the Canadian Army Historical Section. He suggested the title.

Aviation historian Dennis Connell was generous with his time and research for my accounts of Albert Ball and Georges Guynemer. His insight to the events that surrounded *Boelcke's Derelict* were welcome additions.

My thanks to Dennis E. McClendon, author of *Lady Be Good,* who kindly supplied the latest details on the Liberator's lost crew. We would both like to see the final chapter of this mystery written with the recovery of Sergeant Moore's remains—the only member of the crew unaccounted for. I'm also grateful to Dick Baughman, Information Officer of the Air Force Museum, and Henry B. Davis, Jr., Director of the U.S. Army's Quartermaster Museum, for help with this account.

The *Detroit News* furnished accounts of the pioneering flight of Oliver Pacquette and Parker Cramer.

I am grateful to Dr. Douglas H. Robinson, one of the world's foremost airship historians, for clarifying data on the German Zeppelin L-50.

Paul L. Briand, author of *Daughter of the Skies,* supplied sources of information pertaining to the last flight of Amelia Earhart and Fred Noonan.

John Toland, author of *Ships in the Sky,* provided background data on the airship era.

Ted Dealy, former publisher of the *Dallas Morning News,* supplied information on Bill Irwin and Alvin Eichwaldt's flight from his book, *The Last Fool Flight.*

I am especially grateful to those men of the "Jinx Flight," Colonel Thomas Murphy, USAF Retired; William Berkeley; Colonel Francis Thompson, USAF Retired; Lt Howard Darby, USAF Retired; William Frost; Bernard Bennett; CMSgt Adolph Scolavino, USAF Retired; and the late Edward Salley, all of whom freely gave me their time to recount their greatest flying adventure.

C. C. Maher, Development Officer for the Cooma Visitors Centre in New South Wales, provided me with maps, photographs and information for the story of the *Southern Cloud's* disappearance. Ken Williams of the Commonwealth of Australia's Department of Transport, furnished official reports of the crash investigation. Bob Carlin supplied photographs and accounts of the time.

For the story of Bill Lancaster's ordeal in the Sahara, I was generously assisted by Mr. E. P. Wixted, Librarian of the Queensland Museum in Brisbane.

For material related to Glenn Miller, I am especially grateful to Mrs. M. L. Dixie Clerke who had a part in the saga of the famous American bandleader's final flight. C. F. Alan Cass of the Glenn Miller Archives at the University of Colorado at Boulder made important source material available for my use. John Edwards generously gave his time to relate the details of his long search into the wartime mystery of Glenn Miller's disappearance. Lt. Colonel Thomas F. Corrigan, USAF Retired, provided recollections from his experiences of that time. Henry F. Whiston provided maps and background data that helped me put the story in perspective.

Others who assisted me with information were Will D. Parker of the Phillips Petroleum Company; the Lockheed Aircraft Company; Beech Aircraft Corporation; and the Imperial War Museum, London.

My thanks to Robert M. Furman for drawing several of the maps.

Contents

Preface ... vii
Acknowledgments ... ix

PART ONE:
Stories of Two Wars

    1. *The Legion of Lost Warbirds* ... 3
    2. *The Plant that Almost Bombed New York* ... 26
    3. *Derelicts and Ghost Planes* ... 37
    4. *Jinx Flight* ... 50
    5. *Ghost in the Desert* ... 68

PART TWO:
Trailblazers and Adventurers

    6. *Where Is Salomon Andrée?* ... 87
    7. *Somewhere at Sea* ... 116
    8. *The Dole Derby* ... 141
    9. *South to Rio* ... 157
    10. *Flight to World's End* ... 175
    11. *"I'm Going On!"* ... 201
    12. *Amelia Earhart* ... 213

PART THREE:
Famous Controversies

    13. *The Great Floating Palace* ... 243
    14. *"Do You Want to Live Forever?"* ... 259
    15. *The Mystery of Flight 19* ... 282

Index ... 299

Illustrations

Captain Charles Guynemer in his Spad *Vieux Charles*	15
How the completed Poll Bomber would have appeared	34
The wheel of the Poll Giant	34
The *Lady Be Good* in the Libyan Desert	69
The crew of the *Lady Be Good*	71
The *Woolaroc*, winner of the Dole Derby	146
Captain and Mrs. William P. Erwin in the *Dallas Spirit*	153
Paul Redfern and his Stinson Detroiter *Port of Brunswick*	160
The Avro-Fokker airliner *Southern Cloud*	176
Workers sifting through wreckage of the *Southern Cloud*	197
Amelia Earhart and the Lockheed 10 Electra	214

MAPS

Course of the *Lady Be Good*	78
Andrée's route	96
The flight of Paul Redfern	164
Search area for the *Southern Cloud*	185
Lancaster's last flight	205
Amelia Earhart's last flight	217
The Deadly Triangle	292

PART ONE
Stories of Two Wars

1
The Legion of Lost Warbirds

How can a plane plummet into a battleground before the upturned eyes of thousands of entrenched soldiers, smash into their very midst, and disappear without a trace? The fortunes of war are strange indeed if this can happen not once, but several times. It was as though the very heavens—or earth—opened to swallow them and erase all evidence of their fall.

The mysteries of how the missing warbirds met their fates is as puzzling as the disappearances themselves. Bitter arguments still arise, and angry words, charges, and countercharges have ripped the air. In the end, the question marks stand as stubbornly as before.

When a pilot fell behind hostile lines in World War I, enemy Intelligence usually informed the airman's squadron of his death or capture. When possible, they reported his name, rank, and plane number, and included the details of his fall. This was generally done as promptly as channels and conditions permitted, for the enemy expected the same courtesy when one of their pilots was missing. Often the messages were flown to the home base of the victor's squadron and dropped in a cloth streamer. Thus was chivalry practiced among knights of the air.

The Fall of the Eagle of Lille

Max Immelmann flew to early fame and a bizarre death. As a flying companion of Oswald Boelcke, this stiff-backed, proud Saxon was among the first of the German aces. But he was of a different disposition from his warm, popular comrade. Where Boelcke's manner was jovial and friendly, Immelmann was arrogant and conceited. He stood apart from his fellow officers in many ways. He neither drank nor smoked and cared little for the company of even the most charming young ladies. He stuck to a regimen of strict exercise and diet, which excluded meat, and except for a select few he rarely fostered friendships. Max Immelmann was considered rather odd.

The man's strange behavior neither dimmed his fame nor dulled his accomplishments. He was the fatherland's first ace and one of the most dashing figures of World War I. Max Immelmann shot down sixteen enemy planes. He earned his victories when the shooting down of an enemy machine was a feat that automatically led to wide recognition and a decoration or two.

He flew one of the earliest Fokkers, which was equipped with a single, and highly unreliable, machine gun. Those aces who came after him—Richthofen, Voss, Lowenhardt, and Udet—had faster and more dependable planes with twin Spandau guns. Their firepower was five times more effective, and they had many more planes to shoot at.

In July 1915, Anthony Fokker, the Dutch designer who had established aircraft factories in Germany, visited Flying Section 62 to demonstrate his newest machine. It was a small monoplane with an eighty-horsepower rotary engine and a single machine gun synchronized to fire through the propeller arc. Fokker predicted it would clear the skies of Allied airplanes. Immelmann and Fokker hit it off right from the start. Fokker accepted Immelmann's technical knowledge as a near match for his own. Before a week had passed, he promised the fighter pilot a job in his factory "when the war is over."

By August first, Immelmann and Boelcke were each equipped with the deadly monoplane. On that day Immelmann shot down his first plane—the first victory of the war scored by a pilot flying alone. Other victories followed, and when Boelcke was transferred to the Metz sector, Immelmann became the leading ace in the area of Lille. Here he gained most of his victories; here he became known as *der Adler von Lille* (the Eagle of Lille).

In the fall of 1915, Immelmann perfected an aerial maneuver that bears his name—the Immelmann turn. He would pull his tiny monoplane up into a half-loop and, as the machine approached the inverted position, he would suddenly roll it upright. Thus, in a dogfight he would have the two advantages that fighter pilots need badly at close quarters: a gain in altitude and a complete change in direction. Some air historians claim Immelmann never invented the maneuver. Nevertheless, he made it deadly famous.

Victory followed victory; decoration followed decoration. He won the Iron Cross and the coveted *Pour le Mérite*. In June of 1916, Flying Section 62 was ordered to the eastern front, but Immelmann was directed to remain in the west and form his own "chaser" squadron. He was excited by the honor, but he did not live to see the job completed.

On the morning of June eighteenth, he shot down his sixteenth and final victim. Late that afternoon he took off again and flew into a fierce engagement between four Fokkers and seven British machines. It was his last taste of combat. Something happened during that air battle northeast of Douai that has been a dark mystery for over half a century. There are four versions of what happened, each distinctly different from the other, and all provide enough conflicting data to keep the controversy alive for another five decades, and more. But it is unlikely that the mystery of Immelmann's death will ever be solved. All the evidence is in.

As the famous ace made contact with the British fighters, he saw that German antiaircraft had set up a close barrage. They had found the altitude of the dogfight, for shrapnel whistled dangerously close. Possibly the German gunners had mistaken the Fokker for an Allied plane. Immelmann dropped a white signal to tell the men on the ground that they were firing on a friendly aircraft. Then he took after a British machine. A few bursts apparently disabled it, for it started down in a glide. Immelmann decided not to follow it down as was his usual custom, but instead turned back toward the whirling dogfight.

A few seconds later, observers near Annay saw Immelmann's fighter suddenly nose upward. Then it whip-stalled and fell into a diving right-hand spiral, gaining speed all the while. It began to gyrate wildly. Brace wires snapped. At eight-thousand feet the slender fuselage began to whip in the slipstream and a moment later the entire tail section twisted free, leaving the control cables flapping in the rush of air.

The Fokker began to disintegrate. Now the entire rear part of the fuselage ripped free. The wings buckled and finally, with a rumble and a wrench, the rotary engine tore loose from its mounts and dropped like a stone for the remaining five thousand feet. It buried itself in the ground, a twisted, broken mass of steel. Overhead the light tail planes fluttered down in lazy curves, landing a few hundred yards from the main part of the wreckage.

Only one man knew what went wrong, but he was dead.

When Tony Fokker heard the news of Immelmann's death he was upset. The two had been close friends. It was Immelmann who sold the German High Command on several of Fokker's inventions. Fokker was not prepared to blame his friend's death on the failure of one of his airplanes, far from it. He demanded, and was granted, permission to inspect the wreckage. Two days later, to a party of representatives from the German General Staff, he and his sharp-eyed technicians pointed out certain "obvious facts." Of this investigation he later said:

> Immelmann's plane suddenly fell to the ground as he was flying near the German lines. It was first given out that his Fokker fighter had failed in mid-air. This explanation naturally did not satisfy me and, I insisted on examining the remains of the wreck, and establishing the facts of his death.

What I saw convinced me that the fuselage had been shot in two by shrapnel fire. The control wires were cut as by shrapnel, severed ends bent in, not stretched as they would have been in an ordinary crash. The tail of the fuselage was found a considerable distance from the plane itself. As he was flying over the German lines there was a strong opinion in the air force that his comparatively unknown monoplane type, which somewhat resembled the Morane-Saulnier, had been mistaken for a French machine.

Tony Fokker made no bones about defending the soundness of his machines. He insisted that antiaircraft fire—German antiaircraft fire—caused his design to break up in mid-air. But when the nation went into stunned mourning at the news of Immelmann's death, the powers decided it was far better to preserve the image of the ace's invincibility by claiming structural failure of his machine. Oswald Boelcke concurred. Not long before he was killed in an air collision over the lines, he gave this version of the fall:

> Immelmann lost his life by a silly chance. All that is written in the papers about a fight in the air and so on is rot. A bit of his propeller flew off; the jarring tore the bracing wires connecting up with the fuselage, and then broke away.

Admit that Immelmann had been killed in air-to-air combat? Never! It would have been too embarrassing for the High Command to explain to his adoring public.

Now a cloud of doubt hung over the Dutch designer's reputation and Fokker found it highly uncomfortable. He had several heated meetings with the German High Command and finally he and his airplane were exonerated, unofficially, from blame. But many years were to pass before the German people learned there were other accounts of Immelmann's death.

On the British side of the lines an entirely different story of

Max Immelmann's last patrol had been forwarded through channels with an official stamp of approval. As convincingly as Fokker had argued with German officialdom that his plane was shot down by German gunners, a British combat report for the same air fight presented other evidence. The startling thing about this report is that it was published before Immelmann was known to have been killed. Significant points coincided. It stated that on the evening of June eighteenth, Lieutenant G. R. McCubbin, a South African pilot, and Corporal J. H. Waller, observer-gunner, of Squadron Number 25, while flying an F.E.2b, shot down a Fokker monoplane over Annay. Immelmann had crashed at Annay at precisely nine P.M. on this date.

For the fourth account we turn to the man who was Immelmann's flying companion, Corporal Heinemann. He had been above his leader when he saw the strange antics of the Fokker. He claimed that neither a shrapnel burst nor the F.E.'s gun were responsible for Immelmann's death dive. He, too, was curious as to what might have caused the mishap, so he inspected the wreckage, at a different time than had Fokker and his men. His findings definitely did not agree with those of the aircraft's designer. He found two things: (1) the metal tubing of the fuselage had been flattened and pulled before it separated, pointing to tension stress (pulling apart) rather than shrapnel damage, and (2) one blade of the wooden propeller had been cut through by bullets. Halves of bullet holes were found along the line of breakage, which was exactly in the line of fire of the plane's fixed machine gun. To Heinemann, the cause of his commander's death was crystal clear. The mechanical interrupter gear (the synchronizing device that enabled the machine gun to fire only when the propeller blade was out of the way) had temporarily failed. It had happened on other machines before, and now it had happened to Max Immelmann who, according to Heinemann, had been shot down by his own gun.

Who was right? Fokker and his engineers? Boelcke? The

German High Command? The Royal Flying Corps? Corporal Heinemann? All gave vastly different accounts for vastly different reasons.

Choose your own answer; the real one will never be known.

The Mysterious Death of Albert Ball

When it comes to sheer mystery amid a dark and foreboding background, the strange death of Captain Albert Ball on May 7, 1917, is difficult to match. This British ace, who was barely twenty, sent forty-four German planes to earth. He was a popular hero whose death came riding in darkness and mist—a death that set the combatant nations searching for the details. He simply disappeared into a shadowy cloud bank of towering proportions after one of the most bitter and drawn-out air battles of the war.

Albert Ball was the son of Sir Albert Ball, once mayor and alderman of Nottingham. He was a patriotic and deeply religious youth. He paid for his own flying lessons, then joined the Royal Flying Corps in 1916. His early flying skill left much to be desired, but he learned fast and developed into a topnotch fighter pilot. Into his brief lifetime he jam-packed enough flying experiences to suit a dozen combat pilots. Curiously, he didn't record many of his victories; at least eighteen were left unconfirmed.

Normally, the British gave little publicity to even their greatest air fighters, but early in 1916, morale on the home front hit an all-time low. Something had to be done to bolster the sagging spirits. The zeppelins and Gothas were raiding London with deadly regularity, and confidence in the Royal Flying Corps was shaken. Then headquarters came up with the answer: a popular hero. Albert Ball, young, clean-cut, typically British, was their choice. His frequent victories were already in the dispatches. Here, ready-made, was the very man who could keep his senses under a torrent of public adulation.

In October of 1916 he was taken from the front and made an instructor in England. He hated it and tried to pull every string—military, political, and otherwise—to get back to France. But it was not until April of 1917 that he saw action again.

At five-thirty on the afternoon of May seventh, Captain Ball led one of three flights of S.E.5s over the lines. Following him were Flight Leaders Meintjes and Crowe with their flights. Orders were to patrol until nightfall and to ferret out and destroy any German aircraft in the sector.

The weather was foul; a depressing, gloomy day. Rain drizzled intermittently from the cloud layers. Spirits were low and the men were near exhaustion from an unending blur of hectic combat. Ball led "A" Flight, which consisted of himself and two other pilots, along the lines at seven thousand feet. The other planes of "B" and "C" Flights, seven in all, climbed to nine thousand feet. The visibility was so poor that within a few minutes all the flights lost contact with one another.

Suddenly the swirling cloud wall vanished and the three flights found themselves in open sky. They also found themselves smack in the middle of Richthofen's Circus.

The S.E.5s tangled with the enemy Albatroses at a fast clip. Manfred von Richthofen, the circus master, was on leave, but his *Jagdstaffel* was ably commanded that evening by his younger brother, Lothar. In seconds the sky was suddenly crisscrossed with tracer threads, gaudy painted machines, roaring, belching engines, and clattering machine guns. Here and there a widening plume of black smoke fell from the melee to mark another pilot who had come out second best.

The fight reached fever pitch as the thirty machines rolled and twisted in the gathering gloom of nightfall. Most air engagements were brief, but this one showed no sign of letup. It was bitter duel to the death. The planes spread out and covered a widespread area of the rain-drenched sky. Pilots could barely recognize friend from enemy. Spads of

Number 19 Squadron and Sopwith triplanes from Number 8 Naval Squadron piled onto the heap to assist the outnumbered S.E.5s.

Lothar von Richthofen, flying an all-red Albatros with yellow ailerons, signaled his pilots to re-form for another assault. The British fighters broke away to regroup over Arras, and as they separated for a renewed attack they exchanged parting bursts.

Ball was somewhere in that scrap. As Captain Crowe tried to assemble his scattered remnants over Fresnoy, he looked down and sighted his commander at four thousand feet, heading farther into enemy territory. The light was now very poor, and several pilots who found themselves alone, turned for home. Crowe, too, decided the show was over for the day. His goggles had been shot away by a machine-gun bullet, and at that moment five German pursuits appeared above him. He broke off combat.

Crowe glanced down again, this time to see Ball fire two Very flares—the "tallyho" signal. Ball was chasing a lone Albatros. He fired a burst into it and curved away. Crowe dived on the same machine without result, then he too curved away. Ball reappeared, and he and the German went around and around, trading fire in dizzying circles. Crowe turned to make a second pass at the enemy plane, but before he could come within range, both machines were swallowed up by a dark, ominous cloud bank. This was the last ever seen of Albert Ball by a British combatant. Crowe circled about for several minutes, waiting for Ball to emerge, but he never did. Then, with his gasoline running low, Crowe headed for base.

On May eighth, in a German hospital near Lens, a German officer visited a certain Captain Hunter who was an English prisoner recovering from battle wounds. He told Hunter that Ball had been killed. Hunter refused to believe it. The next day the officer returned with Ball's identity disc as proof. He also stated that Ball had been brought down by antiaircraft fire, and his machine was badly smashed.

Three weeks passed before the Germans dropped a note telling of the ace's death. It was delivered to Number 56 Squadron.

> R.F.C. Captain Ball was brought down in a fight in the air on May 7th, by a pilot who was the same order as himself. He was buried at Annoeullen.

The mysteries? There were many.

The credit for downing Ball was given to Lothar von Richthofen when the victory went unclaimed. But Lothar had claimed a Sopwith triplane, not an S.E.5 biplane. Quite a difference. Lothar might well have been the pilot Ball was last seen sparring with, and his confusion with the "triplane" could have been the result of an earlier air fight. In the final analysis, he wasn't able to give a coherent account of what had actually happened.

In the clouds, Ball apparently spiraled on down to a low altitude. When he broke out of the mist in the vicinity of Lens, he was widely separated from his adversary. A German machine-gun crew stationed in a nearby church steeple was one claimant of the Englishman. He had veered close; they had fired, and the machine fell.

The strangest thing of all occurred when French civilians lifted Albert Ball's broken body from the wreckage. They found no bullet wounds on his body. His badly smashed plane verified that it must have fallen from a great height, but although it struck with great force, it did not burn.

Did Captain Ball become disoriented in the cloud while trying to fly blind? Did he stall, fall into a spin, lost control, and plummet to his death, without being hit by gunfire in the air or from the ground? An ironic possibility for a man with the superb coolness to take on any odds.

Air war historian Dennis Connell of Beaumont, Texas, has studied the strange death of Albert Ball and arrived at certain conclusions of his own.

A German officer, Carl Hailer, who was still living in 1969, claimed he viewed Ball's body immediately after the crash. He saw no bullet wounds. He said he saw the S.E.5 emerge from the low cloud, "flying upside down . . . with its propeller stopped" and leaving a trail of black smoke which he considered to be caused by "oil leaking into the cylinders." The officer said the flier's injuries were limited to a broken back, left leg and left foot with some bruises on the left side of the face (which disappeared some hours after death).

In Hailer's opinion, Ball likely became disoriented in the cloud bank and was not aware that he had turned upside down until it was too late to save himself. The officer added, however, that it was possible that Lothar von Richthofen had hit the S.E.5 with "a stray shot." He said he noticed one bullet hole in the breech of Ball's Vickers gun.

All of which advances another theory. One of Ball's fellow pilots said that when he caught sight of Ball's machine the Captain was flying in an odd manner, which led the pilot to believe that Ball "was badly wounded, possibly blinded." Ball, so it's been said, didn't like to wear either a helmet or goggles. Isn't it possible then, that a bullet slamming into the breech of his gun—a scant few inches in front of his face—splintered and lodged metal fragments in one or both eyes, thus blinding him? Unable to see, he crashed. I've known of similar incidents in which persons have had steel fragments driven into their eyes without visible wounds showing. They were very small fragments, of course, but nevertheless quite painful and, although they would not cause permanent blindness, they did impair vision for a time.

Despite Hailer's statement to the contrary, others claimed Ball suffered a head wound. Too, there was the German nurse who wrote to Ball's father to tell him that his son's death was actually caused by a heart attack during the stress of combat. The restrictions of wartime may be responsible for the incomplete reports. Perhaps none of the accounts tell what really happened. A great *perhaps*.

The ace was awarded the Victoria Cross posthumously and his body was buried with military honors in the German war cemetery at Annoeullin. The white marble cross erected after the war by his father can still be seen there.

Borne by Angels

His comrades waited in vain for a message of hope. At twenty-two, he was France's leading ace with fifty-four German machines to his credit.

His name was Georges Guynemer.

Twice rejected by the French Army, his gaunt frame wracked with tuberculosis, this pale little figure with the burning eyes threw himself again and again into the bloodiest part of the fighting. He was shot to earth eight times, but after each crash the steel-nerved champion of France emerged unscathed from the splintered wreckage to mount the skies again in search of the hated Hun.

The phenomenal luck of "Guynemer the Miraculous" ran out on September 11, 1917. The man who had risen to command the famed *Stork Escadrille* took off that day to fly in company with Lieutenant Bozon-Verduraz and Captain Deullin. Over the lines northwest of Ypres, Guynemer sighted a German two-seater Aviatik, and with a quick signal to tell his companions to remain behind as cover and lookout, the ace banked *Vieux Charles,* his favorite Spad, into a dive.

Deullin apparently became detached from the other two at this point. Bozon-Verduraz later said that he last saw his commander plummeting toward the enemy, then found his own hands full as eight German fighters appeared. He lured the attackers away from Guynemer, hoping they would not spot him until the ace could finish his work. After a short encounter, Bozon-Verduraz slipped away from the Germans and returned to the place where he had last seen Guynemer.

The sky was empty and silent.

Bozon-Verduraz searched the front and scanned the sky

U.S. Signal Corps

Captain Charles Guynemer in his Spad *Vieux Charles.*

overhead until his dwindling fuel supply forced him to return to base. Guynemer never returned.

The Storks kept a faithful vigil. Field telephones were busy all along the front, seeking some news of the missing flier. There was no news. The Germans released no victory dispatches for several suspense-filled days, as they assuredly would, had one of their pilots brought the famous *capitaine* down.

More days passed, and reluctantly the French were forced to release the dispatch that told of Guynemer's apparent loss. Then the Allies heard a rumor that a German aviator named Wissemann had killed the Frenchman. The French asked the Germans for confirmation, and finally an official German communiqué listed the ace's death. But it was given as eight A.M. on September tenth, not eight thirty-five A.M. on the eleventh. The Red Cross followed this with a statement that Guynemer had been given a military funeral at Poelcapelle, Flanders.

From this point, the mystery deepens.

Almost a month later, on October 4, Allied infantry took the town but were unable to find Guynemer's grave. Now, the Germans volunteered yet another answer. Guynemer, they claimed, had been brought down south of Poelcapelle Cemetery, but they omitted the details as to how he had been brought down. A surgeon from an infantry battalion, they said, had seen the body and examined it closely. He found that the ace had been shot through the forehead and that the forefinger of his left hand had been shot away. He had suffered a broken arm and a broken leg. As the area was then under heavy Allied artillery fire, the surgeon and his medics were forced to leave the body beside the plane. They, too, the report concluded, were now dead, and the entire area had been churned up by artillery shells. The impression was given that all trace of Guynemer and *Vieux Charles* had been obliterated—scattered and pounded into the Belgian ground by the heavy barrage. The only word that could be reasonably accepted was the hastily scribbled field memo of a methodical German surgeon who was under fire as he wrote it, and who was himself only a few minutes from death.

From extensive research, Dennis Connell came forward with a reasonable explanation as to why the Germans were reluctant to give an early report on the ace's death.

No surgeon examined Guynemer's body. The examination was made by a medical orderly and two privates who were driven away from the crash site by a strafing Allied fighter. They retrieved Guynemer's identification papers, but his name meant nothing to them. It wasn't until the papers were passed on through channels to upper echelons that someone finally realized who Guynemer was. Meanwhile, the German division holding that part of the front where the ace fell was pulled back for a rest that very afternoon. The commander left a request that the replacing division bury the dead Frenchman—but apparently this was not done. By the time German headquarters started making inquiries, Guynemer and his plane had been obliterated by artillery fire. My conclusion is that the Germans, who were very big in the chivalry department, were embarrassed to discover they had let a French national hero come to such a pathetic end so, as a cover-up, they falsified the report that he had been given a military funeral.

Was, then, a body ever buried at Poelcapelle Cemetery in a grave that was never found? If so, whose body was it? And what of Wissemann, the pilot who was credited with the victory? What were his comments? Connell reveals that the Kurt Wissemann who was credited with downing Guynemer was a fighter pilot with Jasta 3. After studying the various fighter pilot claims for that day, German headquarters decided the evidence favored his claim. Wissemann's account said that Guynemer had dived on him, overshot, and that he—Wissemann—swung onto the Frenchman's tail and shot him down with a handful of shots. This was corroborated by Karl Menckhoff, later a famous ace, who was in Jasta 3 and reportedly witnessed Guynemer's fall. When Menckhoff was captured by the French in July of 1918, he told them what had happened. They apparently chose to ignore the story. Wissemann could not confirm it; he had been shot to his death in September of 1917, by 56 Squadron, R.F.C.—a few days after Guynemer fell.

Did Guynemer meet a fate the Germans thought best to hide? Did he die in prison? The wall of silence has never been conclusively breached. To this day, the children of France believe their hero never really died, but that he was borne by angels and "flew so high that he could never come down..."

What Became of the L-50?

Nothing during World War I struck more terror into the English populace than the zeppelin raiders. On one such raid, nature had to do what the island's air defense could not do—bring down the lumbering gray ghosts.

These first air raids by the throbbing cigar-shaped monsters had great psychological impact. In this new type of warfare, they could glide smoothly overhead, out of reach of the highest-flying fighter planes and above effective antiaircraft reach, hover motionless at night and select their targets with deadly precision.

By the fall of 1917, Germany had developed mass formations of zeppelin raiders. One of the largest raids, set for October nineteenth, proved to be the most devastating attack of the war, but it was also the one in which the zeppelins suffered their heaviest losses. It was known as the Silent Raid.

Shortly after noon, eleven Navy zeppelins left their masts at Tondern and nearby coastal bases to rendezvous over England. All afternoon they droned toward the south shore and, by six-thirty in the evening, the first raider crossed the coast at 13,500 feet. Antiaircraft batteries and specially designed rocket volleys reached up for the huge bomber, and when the other airships arrived, the commanders gave orders to rise to safety. The fleet finally leveled off at 16,500 feet.

Darkness fell, a moonless night with black clouds. The airships were well hidden from the probing searchlights that tried to find them in the clouds. In the next few hours the fleet spread out over Hull, Sheffield, and Grimsby, and guided by lights and glowing chimney pots, they dumped their lethal loads and slipped away. The damage was heavy and widespread, and because of poor visibility and dense cloud cover, the zeppelins completely escaped ground fire and pursuit planes. To all appearances, the most successful raid of the war had been accomplished without loss.

But nature has a way of robbing men of ill-gained glory. By midnight, the returning airships discovered that the brisk tailwinds, which had been boosting them homeward, were suddenly growing troublesome. Unknown to the zeppelin commanders, the winds aloft had shifted rapidly in velocity and direction. Up to ten thousand feet, they were still light, but between ten thousand and twenty thousand feet (the zeppelin's cruising altitude), the winds rose to forty miles per hour, and at higher altitudes, over twenty thousand feet, they fast approached gale force.

As the flying machines struggled in the mounting windstorm, their lacy structures creaked and groaned against the turbulence. In the distance, in Flanders Fields, the steady flash of artillery could be seen by the crewmen. At three A.M. the airship navigators were forced to admit that their return flight was not proceeding as planned. They were still over the water when a British destroyer and several land batteries opened up on the armada. The uncomfortably close barrage of antiaircraft fire and rockets caused the zeppelin commanders to jettison ballast and force a quick rise to the rough winds at twenty thousand feet. Now the sputtering diesels gasped for breath in the thin air; the crews suffered from the icy-cold blasts and lack of oxygen. They were no better off there, for the varying winds began to scatter the airships over a wide area.

Six of the airships made it safely back to base, but at daybreak the other five found themselves adrift and helpless over France. One Allied fighter plane spied the L-55 and harassed it up to 25,300 feet before it abandoned the chase. The zeppelin's crew stood up bravely against the sudden rise to new heights; a few became unconscious from trying to work in the rarefied atmosphere, several suffered broken eardrums and blood spurted from their nostrils.

Despite the alcohol in the engines' radiator systems, the coolant mixture froze. So did the water in the ballast tanks. Lieutenant-Commander Hans Fleming became desperate and finally valved gas and jettisoned fuel to land at Tienfort, Germany, safe and sound.

Fleming's companions didn't fare so well. Within minutes of the L-55's landing, the L-44 was shot down by antiaircraft fire as it tried to cross the lines at Luneville. The three remaining zeppelins, the L-45, L-49, and L-50, were still aloft over France until, at eleven twenty-five on the morning of October twentieth, with only an hour's fuel remaining, Captain Koelle of the L-45 ordered a landing in the Durance Valley. As soon as the airship touched lightly on a sandbank in the middle of a small river near Sisterton in the South of France, Koelle set his men to work. In a few minutes they had smashed the engines and punctured the gas bags. Then Koelle ordered his helmsman to fire a signal pistol into the leaking hydrogen. The man obeyed, and within seconds the zeppelin went up in a ball of red-orange fire.

Meanwhile, at Bourbonne-les-Baines, the L-49 was slowly coming down. When it landed, the crewmen stumbled dizzily about, too groggy from the effects of oxygen starvation to get organized. Before the Germans could recover their senses and destroy the airship, French soldiers were on them and the zeppelin was captured intact.

Only the L-50 remained aloft. With his fuel running low, Captain Schwonder turned east, hoping to fly over the French Alps and reach asylum in nearby Switzerland. Although his crewmen wore crude oxygen masks, most of the supply of the life-giving gas had been exhausted by the time the mountains came into sight. Schwonder knew that in the thin air at high altitude, his engine mechanics in their exposed gondolas were falling unconscious. He had ordered the oxygen strictly rationed in the control gondola, and now his officers were becoming drowsy and slow in responding to orders and instructions.

With its six unattended 240-horsepower engines bursting full out, the L-50 headed for the peaks near Dommartin. It almost cleared the first peak, but the helmsman was too weak to move the rudder wheel. Captain Schwonder tried to help him; then, sensing that he too was about to pass out, he grabbed for the power signal and rang STOP ENGINES. Nothing happened. Those mechanics in the engine gondolas who had heard the order were slumped on the engine catwalks, too groggy to move. Schwonder later reported that at this point he tried to ram the airship head-on into the ridge.

Clumsily, the L-50 collided with the peak, but Schwonder

succeeded only in wiping off the control gondola and one aft engine car. Somehow the men in both capsules survived the impact, as did those who regained their senses in time to jump from their gondolas. They crawled painfully from the broken wreckage, stumbled to their feet, and watched the airship strike another mountain. The lightened airship shot upward to twenty-three thousand feet and slowly faded into the fog and mist.

Schwonder and several crewmen survived, spared the unknown fate of the four who drifted away, unconscious, in the pilotless L-50. Later that day the derelict hull floated over the captured crew of the L-45 at Sisterton. As it continued on out to sea, it was chased by French fighter planes that failed to alter its course with destiny. At last report it was far out over the Mediterranean, where it was finally lost to sight.

No trace of the L-50 or its remaining crew were ever found. It had vanished as completely as the *Mary Celeste*. Perhaps it settled into the blue waters of the Mediterranean; perhaps, in the early morning hours of October twenty-first, it slipped unnoticed across North Africa's coast and drifted into the Dark Continent. Perhaps it came to rest somewhere in the vast stretches of the great Sahara. No one knows.

The Ace Who Was No Sportsman

Among the British subjects interned in Turkey when it aligned itself with Germany in World War I, was a tall, spare telephone mechanic. In April of 1915 the Turks released him along with other prisoners they considered too unfit to be of military value to the enemy. The man was over thirty, frail, and in poor health. He had suffered from birth with a badly deformed and almost sightless left eye. The Turks would not have felt so confident had they had any way of knowing that this man was to become Britain's greatest air fighter— Edward Corringham Mannock.

One of the puzzles of Mannock's enlistment is how he

managed to pass the physical examination when he volunteered for the Royal Flying Corps in August of 1916. But pass he did, and started his flight training within a few days. To compensate for his handicapped vision, he had to apply himself to the flying lessons with great diligence. Landings were a nightmare. He spent extra hours in the gun pits, trying to improve his aim. From the moment of his arrival at the front during "Bloody April" of 1917—the month the British suffered their highest air losses—until his death fifteen months later, his victory score never slowed its upward climb.

Major Mannock was a strange, driven man. Old for a fighter pilot (he was in his mid-thirties by then), he became the leading British ace. Only Baron von Richthofen (eighty victories) and Réné Fonck (seventy-five victories) bettered his score of seventy-three victories. His unorthodox cold-blooded behavior toward the German fighting man was the only thing that prevented him from becoming an idol to his men. Although he was a staunch patriot, the British never learned to admire him as they did Bishop, McCudden, Barker, and Ball. But Mannock didn't seem to care. Weird quirks showed themselves in his character. He seemed obsessed, emotionally disturbed. The strain of day-in, day-out combat flying began to show itself after a few months, and Mannock made no secret of his pathological hatred for the Huns. He spoke with genuine pleasure of such things as a doomed rear gunner in a German plane who had "tried frantically to beat out the flames with his bare hands." When he returned from patrols he would sometimes say, "Sizzle, sizzle, sizzle, wonk, woof," which was his way of describing a German plane he'd sent down burning. He collected with relish grisly souvenirs from his kills.

Mannock's fame was greatly dimmed by his treatment of prisoners. His own men swore they saw him swoop down and machine-gun the helpless crews of enemy planes forced down on Allied ground. When ordered to account for his inhuman act, he replied angrily, "Those swine are better off dead—no prisoners for me!" Small wonder he was scorned by fellow officers.

He gave scant attention to what others said or thought about him. To Edward Mannock war meant killing, with no holds barred. Then the pressure took him almost to the breaking point. In May of 1917, he suffered a serious nervous upset. He fought to bring it under control and to all outward appearances he succeeded. Then he began to suffer lapses of memory. His combat reports were frequently vague and incomplete.

In June, nervous tension dealt him another setback. After one landing, his arms and hands shook so badly that he had difficulty controlling them. He was the picture of absolute fatigue. During his last leave in England, he voiced a premonition of death. "I'm coming to the end," he said, "but I don't know whether it will be in the air or not. I don't know . . ." and his voice would trail off.

Back in France, Mannock commanded Number 85 Squadron. On the night of July 25, 1918, he was drinking brandy in the mess. His score now equaled Billy Bishop's—seventy-two planes—and he was well ahead of other British aces. He was unaware that Bishop would never fly operationally again, for the Canadian ace had been appointed to the Air Ministry in London. Mannock could have spent the remainder of the war getting the one victory needed to break the tie, but he actually cared little about being top man. Killing Germans was his only interest.

Although Mannock had no qualms about sending Germans down in flames, he lived in violent dread of going down burning himself. It was his only fear. The idea haunted him. Before he took off on his last flight, he checked the revolver he always carried in his cockpit and remarked casually to Lieutenant Donald Inglis, who was to accompany him, "I'm not going to burn if they get me. I'll blow my brains out first!"

At five-thirty, Mannock and Inglis were airborne in their S.E.5 scouts. Inglis had been instructed to follow closely and do exactly as his leader. In a few short minutes they were whizzing over the German trenches near Merville, thirty feet up and zigzagging furiously to give the ground troops a poor target. Inglis stuck close to the major's tail and hung on for dear life. Suddenly he saw the plane in front wheel and climb. Then Mannock dived on a low-flying two-seater. He opened up with his guns but his speed was too great and he overshot his mark. Inglis followed and shot up the German's fuel tank. The enemy plane staggered, nosed down, and began to burn.

Mannock swept in and crowded the burning machine closely, with Inglis trailing. They were circling dangerously low, an inviting target for ground fire. Mannock continued to ascend around the flaming victim, even as it crashed and burned. Then, after several quick circles over the wreckage, he leveled off and streaked for home.

Without warning, something went wrong.

A flame burst out of the side of Mannock's S.E.5 and trailed an ugly black streak through the air. A second later the entire plane was enveloped in flames. It spun crazily to earth and exploded near its victim. No one will ever know whether the ace had time to make good his boast and use the revolver. Inglis risked a quick circle at twenty feet to try to get a glimpse of his commanding officer, but the plane was burning too fiercely. British antiaircraft gunners at Hazebrouck watched the entire action through binoculars, and confirmed Inglis's report.

There was much speculation as to the cause of Mannock's death. Did he jump, unseen by Inglis, when the flames first burst around him? The most persistent rumor had it that infantry fire struck his fuel tank, but unless the foot soldiers were firing bullets of the tracer or incendiary type (highly unlikely), this possibility is farfetched. Another theory suggested that a final lucky round from the German rear gunner, who *was* firing incendiaries, found its mark in the S.E.5's fuel tank. Finally, it was presumed that as Mannock closely circled the falling wreckage, a fragment from the flaming victim may have lodged in the S.E.5 and set it afire.

We shall never know beyond doubt. Mannock's death

remains as much of an enigma as the man himself. A German officer removed his valuables and personal effects, noted Mannock's brother's address from his wallet, and sent them home through the Red Cross. German troops buried the major nearby and notified the Central Prisoners of War Committee, but the grave could not be found after the war.

The most puzzling thing of all was this: Although it was alleged that Mannock's body was trapped in the blazing aircraft, none of his personal effects showed traces of exposure to fire.

So rests Britain's "Ace of Aces"—somewhere near Merville. Perhaps he found at last the peace his troubled life could not give him. His final patrol was one to remember, for he went out as he had lived—violently.

Mannock was a curious pilot, driven by hatred, a man not easily forgotten when old soldiers talk of the Great War. But in the hearts of his countrymen he never earned the adoration given comrades of lesser fame. In his home town of Canterbury his name is listed without special note on the war monument with others who served their nation.

Mannock's fame, or lack of it, caused many fighter pilots to reflect that humanity is expected of one, even in wartime.

They're still arguing about the fates of Guynemer, Immelmann, Mannock, and Ball, and the disappearance of the L-50. Somewhere, these magnificent skyfighters are having a grand time talking it over.

They're in good company.

2

The Plane That Almost Bombed New York

When the European conflict entered its third year, the United States was poised to intervene. It was then that the German High Command put its stamp of approval to a military plan so fantastic that even today it defies the imagination of the most devoted reader of Jules Verne and H. G. Wells.

When the die was cast, less than a dozen men in Berlin and Potsdam knew the purpose of the venture. Priority war materials, sorely needed for the stepped-up production of German fighter planes, were quietly sidetracked and routed to selected factories throughout the fatherland. Simultaneously, extreme security measures were put into effect by German Intelligence. They did their job so well, obscured and cloaked the activity of this master project so thoroughly, that the details needed to complete the story are still missing.

If all went well, within a few short months the Allied effort would reel under the most daring assault in the short history of aerial warfare. Unlike the pounding of London with mass flights of heavy bombers, this blow would be struck by a single, but very unusual, aircraft.

In the gathering dusk of a heavily guarded aerodrome near Poll on the Rhine, a huge bomber, a *Reisenflugzeug,* was scheduled to be rolled from the depths of its hangar. Antlike figures—the mechanics—would scurry under its huge span. In the belly of the ponderous monster would rest four tons of bombs and enough fuel to feed its ten engines for eighty hours. Its crew, three experienced bomber pilots and two master navigators, would enter the giant and prepare for takeoff.

At navigators' stations along the forward flight deck were certain instruments never before used in aerial navigation. They were sextants and great-circle charts. Bunks, an innovation for an airplane, were provided along the wide aisle. The navigators had been crosstrained to direct the bomber to its target and then to double as bombardiers. No gunners were carried. Gunnery turrets were, in fact, conspicuously absent. Since the target was totally defenseless and incapable of offering ground or aerial defense, they were unnecessary.

When the plane's rumbling engines warmed, fire pots along the length of the runway would be ignited, and this rumbling, roaring colossus, the mightiest plane ever built, would lumber into the air. To avoid the western battlefront it would head northwest over Holland, then into the darkened North Atlantic, where it would set a course for a city thirty-three hundred miles away. Thirty hours later, the bomber would sweep over the unsuspecting metropolis, its bombs tumbling into the crowded streets. The stunned and panic-stricken populace of a city, which considered itself a safe distance from Europe's battlefields, would realize that, without warning, New York City was being bombed by a German plane.

Was this a mad, suicidal, one-way dash by the most daring of Herman Goering's *Luftwaffe* during World War II?

No. This mission was scheduled to occur in the late fall of 1918. Had not the tide of war suddenly turned, this amazing machine could have flown its deadly mission as planned.

The "Poll Giant" was no idle dream of a crackpot engineer. It was very near the flying stage when the fortunes of war brought its construction to a halt.

The Teutonic militarist is no slouch when it comes to the effective use of psychological warfare. History is dotted with his use of fear weapons. He didn't talk about them, he made them and he put them to use. His use of the V1 and V2 weapons against England during World War II was the first mass employment of a morale-shattering medium. The conflagration that followed crippled Britain's war production. Twenty-five years earlier he engineered the devastating zeppelin and Gotha raids over southern England and developed the first use of psychological air warfare.

Americans are prone to smile when told that their cities were slated as German aerial targets in 1918. "America bombed in 1918! Impossible! Why, Hitler couldn't even do it in 1943!"

But Hitler's *Luftwaffe* would have had to penetrate a thirty-five hundred-mile radar net, naval patrols, and flying boats to reach America. Kaiser Wilhelm's *Lufstreitkrafte* would not. His High Command might well have adapted the proved zeppelin, but for some unexplained reason they didn't. Their airship fleet was more than adequate at the time; it consisted of several formidable Navy zeppelins of the L-70 type. One was the L-71, rumored to have been especially designed to bomb New York City. It was later proved that no zeppelin was ever designed for this reason, though airshipmen never questioned the L-71's ability to fly such a mission. This super airship was completed at Friedrichshafen, in the early summer of 1918, as a high-altitude, high-speed bomber. It was 693 feet long, three hundred feet longer than its predecessors that bombed London. Its seven engines totaled 2,030 horsepower. Although most zeppelins droned along at fifty-five to sixty-five miles an hours, the L-71 cruised steadily at one hundred miles an hour and carried fuel for twelve-thousand miles nonstop.

Fortunately, neither the L-71 nor its sister ships made bombing runs on the United States. Instead, the L-71 stayed close to its mooring mast for the duration of the war. In 1919, it was entered as a contender for the ten-thousand-pound prize for the first nonstop Atlantic flight. On the eve of its departure, Captain Lehmann (later of *Hindenburg* fame) was informed it had been removed from the competition by jittery German officials who thought Americans might not be happy to see a wartime zeppelin hovering over New York City. In due course, and in compliance with the Peace Treaty of World War I, the L-71 was delivered to the British as a war prize.

The terrifying London raids were checked in May of 1917, but only after considerable cost to the Germans. The mounting effectiveness of the Home Defense Squadrons was felt as they accounted for ten percent of the seventy-seven airships the enemy lost. Antiaircraft guns, night fighters, and the English weather took their toll of the mighty zeppelins. In that same month, Germany, looking for another weapon that could break the Lion's spirit, found it in the Gothas and Giants—large multiengine bombers capable of carrying more than one thousand pounds of explosives at twelve thousand feet and higher. Unlike the clumsy, unwieldy zeppelins, they were faster, more maneuverable, and deadlier.

The huge flying machines were based at Ghent, 170 miles from London. With clocklike regularity, formations of from fourteen to twenty-two planes began to appear day and night over the uneasy English capital. They ranged so high that fighters and antiaircraft barrages were unable to reach them. Now, even larger numbers of combat-ready planes and pilots were restricted to the island empire, and the Royal Flying Corps in France was feeling the effect. The Flanders battleground could have put the machines to use. Improved antiaircraft weapons and some help from the elements finally forced the German Flying Service to limit their raids to night missions. In late 1917 and early 1918, the German bombers were forced higher. Their bombing accuracy suffered accordingly, but the huge machines created far more havoc than had the earlier zeppelins, and their losses by comparison were slight.

What, precisely, did Germany hope to gain in the bombing of England? Destruction of her war plants? Neutralization of the naval emplacements and training centers? Hardly. Pinpoint bombing of primary military targets was never intended; it was twenty-one years in the future. Although the raids upset war production and cut munitions output by twenty percent, the overall objective was to cripple home morale and divert much-needed fighting machines from the front. To this end, the raids were justified before stiffened aerial resistance spelled *finis* for the great warbirds once thought capable of leveling London.

But when Germany withdrew its London raiders, it had not abandoned the concept of the heavy bomber. The limited success of the London raids, as well as the expected intervention of a productive America, stirred the Central Powers into thinking in parallel terms of psychological warfare. A giant ocean-spanning bomber, a *Reisenflugzeug*, over New York City would shock the Americans into looking to their own defense. This would hold back men and equipment from Europe's battlefields, at least until the great German offensive could smash on to victory.

The story of this great machine and its assigned mission against New York City did not come to light until ten months after the Armistice. And strangely, it wasn't revealed by the defeated Germans. The monstrous, semifinished aircraft was discovered by a group from the International Aeronautical Commission of Control sent to Germany to prepare an inventory of remaining German aircraft. During September of 1919, the group centered its work around Poll on the Rhine. There, in a darkened hangar of an abandoned aerodrome, they uncovered the components of an incredible flying machine. Its design was unlike anything seen before; its proportions were so gargantuan for its day that it defied credibility. The results of the investigation are contained in a little-known report entitled *Ex-German Aerodromes and Materials in Back and Occupied Areas.*

In chapter three of the report, under the heading "Machines of Interest," is a nine-page report subtitled "The Poll or Forsmann Giant." Nowhere in this illustrated document is it claimed that the assembled components were manufactured at Poll. This suggests the huge, heavily guarded airfield was merely the assembly point and that the subassemblies were manufactured elsewhere under strict security and were shipped secretly to the Rhine aerodrome. The parts contained no manufacturers' marks, nor could the workmanship, praised as "good" by the commission, be traced to any aircraft builder. Although the parent company behind Forsmann's construction of the huge bomber was never determined, some investigators surmised that he was furnished facilities and assistance by Mannesmann Werke, a manufacturer of tube and steel fabrications. The Deutsche Bank was undoubtedly the firm that provided the financing.

Several parts—landing gear, control surfaces, bomb releases—were never found. Nevertheless, after a careful analysis of the components taken from the gloomy hangar, the commission decided the machine was under construction as a "heavy-bombing, long-distance machine, alleged to have been intended to bomb New York."

The word *alleged* invites controversy as to the plane's purpose. Allied chiefs would have considered the planned raid suicidal as well as highly improbable, but more than once they had underestimated the ingenuity of German aircraft design and performance and paid the price over the western front. If anyone has since found another purpose for the big bomber, he has failed to bring it forward. The plane was definitely not a research and development project. Germany could not afford the luxury of diverting skilled craftsmen, critical materials, and valuable time on an off-track military venture that expensive. It was not intended to bomb London or industrial France; the Gothas and Giants were adequate for that. To what possible use could this machine have been put, designed with a nonstop six-thousand-mile range, other than to raid New York City?

The commission estimated the "Forsmann Giant" would have carried fuel for eighty hours of flight at a crusing speed of 120 miles an hour. Sketches attached to the report show an amazing clarity in the plane's design. The skilled investigators who reported these findings were fully aware their words and figures would be carefully read by their supervisors in government and military positions; their accuracy could easily be checked. But the report was never questioned. The sober analysis of the contents of Poll aerodrome hangar was accepted as fact. The Poll bomber did exist.

It was a triplane, with its upper and lower wings spanning 102 feet. Its center plane measured 165 feet. On this wing thirty-three-foot ailerons, which were never found, would have been fitted. The wings had a standard chord (width) of twenty-two feet, with a well-cambered (curved) airfoil to compromise between load carrying and speed. At its thickest part the airfoil measured one foot. There were several equally spaced compression struts that extended five inches below the lower wing surfaces. All the wings had two main spars of sturdy box construction and each one extended the length of the wing. The front spar was unusually close to the wing's leading edge, so for maximum strength, all main lifting surfaces were covered with three-ply veneer, over which a thin sheet of muslin had been glued with varnish. Each wing was separated by eighteen feet. Once assembled, the aircraft's top wing would have been forty-five feet above the ground.

There appeared to be some similarity in design between an earlier bomber, the Schutte-Lanze R1 of 1918, which was designed to be powered with six Basse u. Selve or Maybach engines of three hundred horsepower each. But the "Forsmann Giant" would take ten of these engines arranged back to back. They would have been supported in an unusual pyramid-type strut arrangement to hold the engine bearers on steel-plate fittings and tubing.

Almost as novel as the aircraft's size was its fuselage construction. It measured nine feet and three inches at maximum width, and the square structure was, like the wings, entirely of wood covered with three-ply veneer. It tapered smoothly to a flat wedge at the rear and converged bluntly at the nose. The main members were four longerons and a number of cable-braced members.

No cables, however, crossed the fuselage interior. This was an unobstructed passage for almost all of its 150 feet. There was an unusual raised walking platform running the length of the fuselage, under the forward part of which, it was believed, the bombs would have been carried internally. Eight square windows lined each side of the fuselage. Unlike most crude bombers of the early helmet-and-goggles days, there were no open cockpits; all occupants were enclosed.

The stablizing tail plane, only one side of which was found, was nine inches thick and twenty-eight feet long, plywood covered. The pitch control of the cumbersome machine was of a highly stable design but would probably have created a problem in flight, since both the leading and trailing edges of this normally rigid part were moveable. When operated simultaneously with the elevators, only a slight control pressure would have been needed to put the plane in a climb or dive. The vertical fin and rudder were never found.

As far as we know today, only two parts of the machine remain. A section of the giant fuselage is stored in the Imperial War Museum in London, and is not on display. The other part is one of the wooden wheels taken from the Poll and placed among the museum's aeronautical exhibits. The huge disc measures ninety-nine inches across its center. The laminated beech rim is ten inches wide. There is some doubt that the wheel is complete; officials of the museum suspect that, even with three other similar wheels, it probably could not have supported the heavy aircraft. This hardly disproves its intended use, for the plane was not meant to withstand continuous abuse on rough fields. After its test flight, there would be only one takeoff and, eighty hours later, one landing. This is all that would have been required of the wooden monster. It is likely the gear would have been dropped after takeoff to reduce drag. Once back at Poll, the plane could have been bellied in, or it could have been deliberately ditched at a predetermined spot in the Atlantic where a waiting U-boat would pick up the crew.

Forsmann (whose name has been alternately spelled

Imperial War Museum

This is how the completed Poll bomber would have appeared.

Below: One of the wooden wheels of the Poll bomber, now on display in the Imperial War Museum, London.

Forstmann and Forssman in reports), the intrepid Swede believed to have designed the plane, first worked in Russia as an aeronaut from 1910 to 1912, where he developed lighter-than-air machines. He came to Germany, where he built a small monoplane for Prince Frederich Sigismund. At the outbreak of the war in 1914, he joined the Siemans-Schukert Werke in Berlin. There he designed a number of large aircraft, among them the S.S.W. Giant, a biplane of seventy-two-foot wingspan and a gross weight of five tons. Production models of the Giant bombed London with the Gothas, and one of Forsmann's similar designs carried sixty passengers, a feat in those days. A designer with imagination, he was dramatically innovative. Among his projects was a one-man submarine and a "shell-proof" tank. But the crowning feat of his unconventional mind was the Poll triplane, which he likely conceived late in 1916.

There was nothing to show that work on the project was interrupted before the Armistice. A small, one-page item that appeared seven months earlier in *The New York Times* (April 23, 1918) gives a clue to indicate the Poll bomber was under construction then. It also points out a minor, though distorted, break in German security that leaked through neutral Holland and into the United States. The title of the news item may have been substantially correct, and perhaps only the details were garbled.

New York Times, April 23, 1918
TALKING OF POSSIBLE AIR RAID ON NEW YORK
Special Submarines to Bring Planes Within Reach Being Built, says German paper . . .
Special Cable to *The New York Times.*
THE HAGUE, April 22nd—The *Vossische Zeitung* alleges that it learned indirectly from Paris that the American coast is patrolled by water-planes.

The paper says that Major Havers declares that an air raid on New York is not only possible, but probable, and that special submarines are being built in Germany to carry airplanes which can be dismantled. Each airplane would then drop 100 kilos (2,200 pounds) of explosives on the roofs of New York and would even penetrate 450 kilometres inland.

Americans gave the news item scant notice. A few were amused by the idea. New York bombed? Preposterous!

True, Kaiser Bill's special plane-carrying submarines did not materialize, but this did not preclude another type of construction going on at a secret base in Germany. This was not at the navy pens of Kiel or the humming inland factories, but in the tranquil, picturesque valley of the Rhine. Here, where the world's finest wines are made, was found the hopes of the German High Command—a death-dealing warbird in the making. A flying machine of unbelievable proportions, designed to perform an impossible task—the bombing of New York City. Without a doubt this was Germany's secret weapon of World War I.

When Germany lost the war, work at Poll stopped. Riggers and fitters, carpenters and fabric workers scattered to their homes. The airfield was abandoned and the hangar doors were sealed until the team from IACC opened them ten months later. By then, information was hard to get and the trail was growing cold.

We may never know how close Germany came to getting this B-19 of its day airborne on its mission. All we can surmise is that, given a few more weeks, perhaps a month at the most, World War I's most daring long-range bombing raid might well have come to America's shores, on schedule, on target.

3

Derelicts and Ghost Planes

During World War I, two Australians flew their riddled aircraft on a mission over the German lines. They were dead, but they brought their aircraft home. In 1941, from the great battleground over England, fighter pilots reported a ghost plane that sent German raiders flaming to earth. And more than one squadron of RAF pilots watched a specter escort hover nearby during the height of the London blitz.

Tales of strange apparitions of the skies, corpses at the controls, phantom planes, their mysterious appearances and even stranger disappearances, have been received with skepticism. But despite the contention that "supernatural" sightings are to be expected under severe wartime stress, those who saw—believe.

Boelcke's Derelict

One of the most grisly occurrences of the first air war was "Boelcke's Derelict." It was named for the German ace who allegedly discovered it.

On a crisp September morning in 1916, Captain Oswald Boelcke, who was then the Kaiser's top ace, led five of his ablest warbirds homeward after a successful dawn patrol. Among his pilots that morning was Baron Manfred von Richthofen, the man destined to become the highest scoring airman of the war. Over Armentières, as the tight formation swung close to a towering cloud bank, a British reconnaissance plane abruptly burst from the vapor. It headed under full power directly for the German formation. The squadron scattered, narrowly escaping collision. Boelcke was quick to recover his senses and signal for the attack. One by one his squadron mates darted toward the two-seater and poured round after round of machine-gun fire into it.

Strangely, the British crewmen held their fire. They took no evasive action, but continued to drone monotonously onward in a wide circle to the left.

Perplexed by the strange behavior of the plane and the failure of his finest pilots to down it, Boelcke signaled his men aside, centered the enemy machine squarely in his gunsight, swooped, and poured a long burst of lead into the cockpits.

Unwavering, the two-seater flew on.

Boelcke stared in disbelief. His withering attack apparently had had no effect on the machine. He saw his flaming tracers penetrate the plane; saw the gunner crouched over his weapon. Cautiously, the ace inched his speedy Albatros pursuit closer, ready for an English trick. When only a few yards separated them, he banked his machine and peered into the open cockpits.

The ghastly sight across those few yards of space chilled the marrow of even his hardened air fighter, for there, strapped bolt upright in the wind, were the death-stiffened bodies of the pilot and observer. Their riddled corpses were smeared with blood, their sightless eyes glazed, staring into nothingness.

They were riding a hearse of the sky.

Boelcke escorted the derelict for several minutes, then, with a dip of his wings, he touched his forehead in a final salute and rejoined his waiting pilots. They returned home, leaving the plane to continue its journey to an unknown destination. Boelcke recorded the incident in his combat report but refused credit for the kill. He suspected that the men were already dead when their plane departed from the cloud bank. He was correct, although he never learned of the events that preceded their meeting over the western front. A month later Boelcke was dead.

At dawn that day, two Australian airmen of Number 3 Squadron, Australian Flying Corps, Lieutenant J. L. Sandy and Sergeant F. L. Hughes, pilot and wireless operator, had taken off in their R.E.8 for no-man's-land. Their mission was to direct a ground mortar battery to knock out a German rear position. No sooner had they begun their work in directing artillery fire than they were jumped by a squadron of German pursuits. A mile distant, two other Australian airmen in another recon plane saw their plight and flew to their rescue. One of the Germans was downed before the battle ended. As the rescuers turned away, they waved to the artillery plane which, to all appearances, was back at work. Their gesture, however, was not returned.

A few minutes later, Boelcke met the plane when it popped from the cloud bank. After a vicious attack, he left it to seek its own end.

The battered aircraft lumbered on until, forty minutes later, over St. Pol, its engine sputtered from lack of fuel and became silent. As though by ghostly hands, the derelict glided smoothly to a safe landing in an open field. French soldiers rushed to the plane. They found the bodies of Sandy and Hughes, and transported them to a nearby hospital where an immediate autopsy was made. The findings: both men had been killed instantly by a single bullet, an armor-piercing projectile that had sliced through Hughes's left lung and lodged in the base of Sandy's brain. They had been dead more than an hour when the plane landed.

The most inexplicable part of the incident was this: Of the hundreds of bullet holes in the plane, made by more than a dozen pursuits in the two air engagements, not one bullet had struck a vital part of the propeller or engine. The gas tanks had not been pierced, nor was any part of the plane's control mechanism damaged.

The aircraft had penetrated deep into enemy-held ground, but the wind, blowing from the northeast that morning, drifted the plane back over home ground.

Questions arose as to another, similar, hearse of the clouds. Some reports said that "Boelcke's Derelict" actually appeared on September 27, 1916, when the German flier poured burst after burst into a single-seater Martinsyde G 100, piloted by Second Lieutenant S. Dendrino of Nr 27 Squadron, R.F.C. Curious as to why the plane continued to fly doggedly in circles instead of falling, Boelcke ventured closer and saw the pilot was dead in the cockpit, slumped over the controls in such a way as to jam them. Boelcke wrote, "I left the plane to its fate," and on returning to base, when he heard that another pilot had put in a claim for a machine with the same identifying number, he suggested to the staff officer that "the victory should not be credited to anyone" so as to avoid "doing either of us an injustice."

The incident involving Sandy and Hughes was verified as having happened in December of 1917—a derelict of the sky that sped onward with its lifeless crew to the last drop of gasoline and the final kick of its engine. It was a flying hearse with Death at the controls, a warplane that stubbornly refused to be downed by the best of the kaiser's fliers.

The Corpse in the P-40

A year after Pearl Harbor, the Chinese Warning Net on Kienow Airfield alerted American pilots of the China Task Force that an unidentified P-40 was approaching the mainland from Japanese-occupied Formosa. At the operations room, Colonel Robert E. Scott, Jr.—later Brigadier General USAF—a squadron section leader, followed the colored flags being pinned to a map to show the course of the oncoming fighter. It was coming in low, headed directly for the field.

Colonel Scott pondered. It would be dark in an hour and the weather was poor. Wet, gray clouds hung low at five hundred feet. The Japanese never ventured this far inland in bad weather, and especially not this late in the day. Even in good weather they rarely made a raid when they had to return to their bases after dark. And there was never just *one* plane.

The officer decided not to take a chance; it could be a trick—a captured P-40 used as a decoy. He and another pilot, Costello, were soon airborne in their shark-nosed pursuits to check out the strange intruder. Ten minutes from the field they sighted their target two hundred feet below. Scott radioed Kienow, "Contact gained with a single ship," then he and his wing man maneuvered for position. As they curved overhead Scott saw a strange sight. The P-40 was an old B model *Tomahawk*, the type used at Pearl Harbor a year earlier. Even the insignia was obsolete—a blue disc with a white star and a red center. The plane was badly pockmarked with dozens of bullet holes.

The two fighter pilots flashed recognition signals. No response. They dived together and opened fire simultaneously. Caught in the crossfire of a hundred rounds that thudded through the metal wings and fuselage, the old Curtiss shuddered. The plane tipped only slightly, recovered, and droned on. It made no evasive move.

With Costello to cover him, Scott edged in below the ship and slowly narrowed the distance until he was only a few yards away. He could hardly believe his eyes. The cockpit had been badly shot up, the fuselage was a sieve, the right aileron was gone, and one wing was shorter than the other where part of its tip was blasted away. Staring up at the plane's underside, he saw something else. It had no landing gear or wheels. The wells in the wing, into which the wheels normally retracted, were empty. *How had it taken off?*

Scott radioed his base again, to report the details. *"I don't know what's going on up here. We've intercepted the plane and fired on it, and it's still headed straight for the field—but I don't think he'll make it. He's shot to pieces."* Gently, Scott applied engine power and eased up alongside the phantom plane until he could look directly across the space and into the cockpit. Costello inched up on the other side.

The colonel could see the pilot clearly now, sitting upright, long hair, with a pallid, bloody face. Later he said the mystery pilot must have sensed the nearness of another plane, for he thought he turned his head slightly, raised his hand in a weak signal—then dropped it limply. But Costello told him he had only imagined it; the man was dead—long dead.

Ragged festoons of gray clouds hung from the overcast and, with little warning, all three planes flew into the mist. When Scott and Costello emerged, the ghost ship had vanished. It was the last they saw of it in the air. Then, seconds later, Colonel Scott caught sight of it below, as it struck the ground in a rice paddy. It flipped quickly on its back; there was the flash of an explosion. Darkness was fast closing in, so he tried to fix the location in his mind as they turned back to Kienow.

After landing, he and a doctor drove a truck through most of the night in an unsuccessful search of the rice paddies for the wrecked plane. In the morning, however, a report from the Chinese Warning Net pinpointed the wreckage. Scott and Costello, with the doctor and two engineering men, hurried to the location. They found the remains of the plane high on one of the hills that juts into the clouds that hang over eastern China. Scott saw that the plane had been riddled through from its nose to its tail; from wing tip to wing tip. Gunfire had come from below and above, from behind and in front, revealing that Japanese planes as well as enemy ground fire had damaged the plane. But how, the men puzzled, could the pilot have lived to fly the plane as far as it must have come?

The P-40 had struck the ground in a near-landing position, but it's propeller had caught in the soft mud of the paddy and it nosed over and burned. The question of the missing landing gear came up again. Where was it? How had the plane taken off?

The pilot was dead, his body burned and mangled almost beyond recognition. There was no identification tag and the doctor tried, unsuccessfully, to take fingerprints. The pilot's leather jacket was reasonably intact and in its pockets they found several letters, some scorched, some intact. These

DERELICTS AND GHOST PLANES 43

were mailed to the addresses on the envelopes. The men also found a notebook-diary, partly destroyed, but sufficiently intact to help them piece together—with some conjecture—an account of where the P-40 had come from and what its mission had been. The diary, and tireless research over three decades by Massachusetts author Curt Norris, brought to light a story that would rival a Christopher Wren novel.

In 1945, Norris was on Bataan with the Army Air Force, where he first heard the story of the mystery pilot and his comrades. He began his research there.

The phantom P-40 had come from the island of Mindanao in the Philippines. The Americn pilot had flown there from Bataan on April 7, 1942, and joined a small group of airplane mechanics. When Bataan fell on April ninth, all contact was lost with the scattered American forces.

The Americans—nineteen in all—refused to give up, even after Brigadier General Sharp called for their surrender when Corregidor fell in May. At a hidden airfield in the center of the island they formed a pocket of resistance, cannibalizing wrecked P-40s to keep at least one of them flying. With the plane, the pilot harassed Japanese airfields and shipping with sudden sneak attacks for months. On one mission he flew four hundred miles to strike at Manila—and returned.

The raids ended when the only flyable P-40 damaged its landing gear and propeller during a belly landing. The mechanics straightened the propeller and salvaged parts from other planes, but there was no serviceable landing gear among them. Then someone thought of making long skids from three-inch bamboo, bracing them from the wheel wells to the tail. They could be designed to be released by a pull wire once the plane was airborne. The idea worked and, with the radio and inverter removed from the fuselage compartment, external fuel tanks and a fifty-gallon reserve tank were installed that gave the small fighter a range of thirteen hundred miles—with favorable winds. From a nearby crash-landed B-17, the mechanics drained enough gasoline to fill the P-40's tanks, and the pilot was ready for his long-distance dash to the American Volunteer Group in China—and another chance to strike the enemy.

Each of the men who remained behind gave the pilot a letter to mail to family and loved ones if he reached China safely. It was those letters—the ones that survived the crash—that were mailed home. Any one of the letters from the eighteen Americans left behind could tell the whole story today.

The plane's skids were greased with engine oil to reduce friction as they slipped over the Cogan grass and, on December 7, 1942—one year to the day after Pearl Harbor—the P-40 pilot roared off. Five hours later (as verified by Japanese records) he passed over Formosa and, although the plane withstood withering ground fire and repeated air attacks, the pilot was mortally struck. Whether he died then, or somewhere over the Formosa Strait, will never be known. But seven hours later the pilotless fighter plane crossed the China coastline and soon encountered Scott and Costello.

Curt Norris is firm in his belief that the identity of the unknown pilot is within reach. And he believes some of the letters that survived the crash were mailed to relatives in Massachusetts. If so, do they still lie in a bureau drawer or a box in the attic—or are they remembered, vaguely perhaps, by a widow, a parent, or a one-time sweetheart?

The name of the mysterious pilot from Mindanao may yet come to light, for his imprint on America's first Pacific air war was strong. Memories may yet stir; incidents believed long forgotten may be remembered. The answer could surface soon—or never. Curt Norris thinks the pilot deserves better than a forgotten Chinese grave. So do many of us.

Phantom Fighters of the Blitz

If it is difficult for the average person to accept a phantasm of the dead, how would one go about asking him to believe an apparition of a house or a sailing vessel—or an airplane? Ghosts, it appears, are not limited to the deceased, they may be "things" as well. Machines guided by the personalities of men who persist in hovering near. Of such is the next story.

During the early struggle for European air supremacy in World War II, tales of ghost planes that haunted the battle skies were whispered about in the squadron ready rooms and flight-line dugouts. The "White Angel of Warsaw" was one of these. It preceded German bombers minutes before they arrived over the Polish capital. It streaked silently from the west, shimmering with an eerie luminescence even in bright daylight. The specter plane would circle the city once, then disappear. After Warsaw fell to the advancing German armies, it was never sighted again.

There are still a few ex-RAF pilots who remember the legend of "Old Willie" and some who swear they saw him. He was supposed to have been the ghost of a World War I pilot named Henshaw, a Canadian. Henshaw had an insatiable desire to shoot down a German plane, but in his eagerness he grew careless. On his very first encounter with an enemy machine, he was wounded, lost control of his plane, and crashed between the lines. For two days, without food, he was forced to lie in a water-filled shell hole and nurse his wound as best he could while a heavy battle raged. He was finally able to inch his way back to friendly lines. His wound meant the end of the war for him and he was furloughed home. But until the day he died in 1929, shooting down a *Boche* remained a driving obsession.

A youthful RAF pilot who claimed to have seen "Old Willie," recalled the following:

When the Blitz began, we really had our hands full. During one night patrol our squadron got quite a surprise. We were preparing to attack a large formation of Heinkels when we noticed another plane in our formation. It was British all right. Our squadron leader saw the cockade insignia, but it wasn't the latest-type machine by any means. We tried to signal, but there was no response. Then our leader recognized it as an old Canadian biplane that was somehow managing to keep up with our fast pursuits.

Suddenly it peeled off our formation and screamed straight for the two lead Heinkels. They saw it coming, too, and they swerved in their tight formation, collided, and went down in flames. The Canuck pilot veered over toward us, waved a snappy "thumbs up" and simply disappeared into the mist. He's been seen at night many times by other RAF pilots. They say he always uses the same trick—diving straight for the enemy planes until they collide, or unnerving their pilots until they get careless and become an easy mark for our lads.

While "Old Willie" was the self-appointed protector for the Fighter Command, its counterpart, the "Hot One," looked after the boys of the Bomber Command. This specter ship, a Handley-Page bomber, was guided by invisible hands over enemy territory, on the long-range bombing runs to Berlin. It too had a specialty of its own.

According to the pilot of a Whitley bomber who was flying over "Hellfire Corner" one night:

This huge Handley-Page swooped down on us from above, traveling at a terrific rate of speed. It was brightly lighted and I could see there was no one at the controls.

It dived past and quickly outdistanced us. Suddenly the Huns let loose with a great barrage of flak from hidden antiaircraft batteries on the ground. The bomber was caught squarely in the middle of bursting shrapnel. I saw it in time and turned away.

The bomber must have been hit in a thousand places, but it pulled away ahead of us, climbed steeply into the clouds, and disappeared. If it hadn't attracted the fire from those concealed batteries, the German gunners would have caught us in a trap.

Other pilots have sworn that the timely appearance of the "Hot One" had saved their mission as they carried their bombloads to enemy targets. So, to the venerable and respected ghosts of England's historic castles and manors, has been added a touch of modern warfare—planes of vengeance.

Riddle of the Berserk Blimp

America's involvement in the war with the Axis brought about this bizarre air event—a sky derelict of a different sort.

In the tense months that followed Pearl Harbor, the West Coast of the United States bristled in anticipation of a second sneak Japanese attack. The main concern of the U.S. Navy was that an enemy submarine would slip into San Francisco Bay to further cripple the Pacific fleet. Navy patrol planes and surface vessels were spread thin and had to be assisted by scouting blimps, nonrigid helium-filled bags powered with aircraft engines. They were slow and ponderous, but with a load of deadly depth charges they could hover over a submerged undersea boat and blast it with pinpoint accuracy.

On a certain morning in 1942, at a few minutes past six, Blimp L-8 lifted from its mooring on Treasure Island and headed out to sea. The weather was good, and two experienced airshipmen aboard, Lieutenant Cody and Ensign Adams, expected their mission to be another routine sweep of the approaches to the bay.

The L-8 droned on as the officers scanned the Pacific for signs of unusual activity. They made routine radio contacts and checked on several fishing boats. Then, at seven-fifty, Lieutenant Cody radioed Treasure Island that they had found a large oil slick below.

"Looks like there might be a Jap sub waiting around the channel," he reported. "I'm taking the ship down to three hundred feet for a closer look."

When the captains of two nearby fishing trawlers saw the

blimp circling lower, they knew it was on the track of a sub. They gave brisk orders to drag in their nets and retire to a safe distance, where their hulls would not be damaged by the concussion of an exploding depth bomb. Two armed patrol boats, alerted by radio, also paused at a distance to await the result of Lieutenant Cody's investigation.

The airship turned slowly and droned over the oil slick, but instead of releasing a depth charge, it hovered momentarily, then ballooned upward to disappear in a scattered cloud layer.

About ten o'clock, several fishermen casting nets along the beach near the Coast Artillery Patrol Station looked up to see the L-8 drifting toward them. When the gondola touched the beach, they rushed forward and grasped the draglines to hold down the semibuoyant airship. Their efforts were futile; the blimp dragged them roughly along the shoreline for a hundred yards. Before it finally shook them loose, they looked through the open door of the gondola and saw there was no one aboard. The L-8 was a derelict of the air.

As the huge gas bag raked itself awkwardly along a cliff that bordered the beach, one of its three-hundred-pound depth charges jolted free and plunged into the earth beside a highway. The sudden release of weight gave the blimp new buoyancy. Again it shot upward and away. Thirty minutes later it settled clumsily into the streets of Daly City on the outskirts of San Francisco.

The Navy immediately ordered a salvage crew to the scene and the now-limp gas envelope was emptied, disassembled from the gondola, and transported back to the base for a thorough examination.

What caused the strange and unexpected antics of the L-8? Where were Cody and Adams? Why did the airship ascend suddenly into the cloud layer?

A careful inspection showed all equipment in the control gondola to be in place. Nothing had been damaged. The parachutes were neatly stowed in their racks; the rubber life raft was in its proper location. One investigator noted that the airship's two yellow life jackets were missing, but this was normal. All crewmen were required to wear them on flights over the water.

Then came a surprising discovery. Evidently the blimp had not touched down in the ocean at any time after it departed from Treasure Island. No water was found in the space beneath the deck of the gondola. Did Cody and Adams fall into the ocean as they circled the oil slick at low altitude? Was it the sudden loss of weight that caused the blimp to balloon upward? Apparently not. There was not one member of the four boat crews on the area who had seen bodies fall into the water, and the brilliant-yellow life jackets as well as the splash on impact would have been too obvious to be overlooked. A sea and air search conducted that afternoon and in the days that followed failed to turn up either life jackets or bodies, and the strange case of the berserk blimp and its missing officers is still no nearer to a solution than it was in 1942.

Dead men at the controls. Phantom pilots on phantom wings. Spectral escorts of the night, haunting the skies where once they fought and died. In air wars of the future, will they return?

4
Jinx Flight

From a vantage point in 1966, the former members of a certain B-24 crew can look back on a hoodooed bomber sortie of World War II and agree on one thing—there were no preflight forebodings to that mission.

"We were too young to have premonitions, I guess," reflected Colonel Thomas L. Murphy, who had copiloted the jinxed bomber as a second lieutenant thirty-eight years earlier. "From all the things that happened to me over there, I think I must have been the jinx. I graduated from flying school on Friday the thirteenth!"

Scattered now from Europe to the Pacific, seven of the original nine-man crew are hale and hearty today. And each retains a vivid recollection of four suspense-packed hours shared over the Bay of Bengal as one flight system after another failed on their homeward-limping Liberator. Sergeant Ed Cunningham, a reporter for the wartime *Yank,* said of their fantastic survival, "They played tag with borrowed time so often, the law of averages is in grave danger of being repealed." But when the war ended, five of the men remained with the Air Force. All were highly decorated; some were awarded the Distinguished Flying Cross and the Air Medal.

The enlisted members of the crew began training as a team at Salina, Kansas, early in 1942. They were M/Sgt. Howard C. Darby, bombardier; T/Sgt William O. Frost, engineer; S/Sgt John E. Craigie, radio operator; S/Sgt Bernard L. Bennett, tail gunner; S/Sgt Adolph R. Scolavino, belly gunner; and Sergeant Edward M. Salley, waist gunner. By September, the group was in east-central India where they quickly shaped into a seasoned combat team of the Tenth Air Force based at Allahabad. Of their quarters in the ancient holy city, Salley said, "We lived in the McPherson Barracks, once the home of the famed Bengal Lancers. The buildings were stone, two-story, and very large. We slept on wood beds with rope stretched across them to support straw mattresses. The field was fairly modern, with concrete runways."

Their B-24D carried a picture of a green shamrock on its blunt nose. It belonged to Flight 1 of the 436th Heavy Bombardment Squadron.

Three days before a raid was scheduled to come off on Rangoon's railroad marshaling yards and shipping docks, a briefing was held for the crews that would take part. Six aircraft were assigned the mission, with its flights directed to have a five-minute separation. Cruising at twenty-two thousand feet with an airspeed of 220 miles per hour, they were timed to strike the target at dusk or shortly thereafter. In Flight 1, the shamrock-embellished bomber would lead with First Lieutenant William R. Berkeley as pilot, Second Lieutenant Thomas L. Murphy as copilot, and First Lieutenant Francis N. Thompson as navigator. Each B-24 would carry its capacity load—twelve thousand pounds of demolition bombs. The alternate target was the Japanese airfield north of the city.

William Berkeley, who retired as a lieutenant colonel in 1960, recalls that at about two P.M. on the afternoon of October twenty-ninth, he lifted the heavily-loaded bomber from the strip at Allahabad and headed southeast on the thousand-mile run to southern Burma. In the right-hand seat was the man who became a colonel in the Strategic Air Command, and the Director of Safety at Barksdale Air Force Base, Thomas L. Murphy. Murphy flew thirty World War II missions in B-24s and B-25s.

"The mission was to involve five hours of daylight flying

and four hours at night," Murphy recalls. "It was led by a 'pink elephant'—a B-24 recently transferred from the African desert theater. I don't remember the command colonel's name, but he was supposed to show us how to lead a bombing raid. As it turned out, monsoon weather and his unfamiliarity with weather-penetration procedures for our formations broke up the flight before we reached the target."

Despite the gauntlet of monsoon thunderstorms and the disturbing fact that Berkeley was forced to pull near maximum power to keep up with the other flight, Lieutenant Thompson zeroed the bombers over the target as the sky was darkening into nightfall. It was then that the plague of unexplained electrical and mechanical failures began—failures that ultimately forced nine men to spill into the pitch-black sky a mile and a half over Bengal Province.

M/Sgt. Darby was lead bombadier of his flight. Today he is retired from the United States Forest Service in California.

"We were told the weather would be good over the target," Darby said, in looking back to that memorable raid, "and the forecast proved to be generally accurate. I was preparing for the bomb run when I smelled something burning. It was my chute. Thompson, who was nearby, pulled it away from a heater, just before I dropped the load of thousand pounders. He put the fire out with an extinguisher."

At takeoff, Darby's parachute was inadvertently laid over the heater that was used to defrost and warm the Plexiglas nose section. The scorched pack meant one man would be short if the crew had to abandon the plane, and a familiar joke then making the rounds among the fliers suddenly became a deadly serious matter. It went something like this:

> A bomber crew was bringing their plane home after a mission when they were hit by enemy *ack-ack*. The engines began to smoke and misfire as the plane lost altitude rapidly.
>
> "Anyone know how to pray?" the pilot asked over the intercom.

"I do, sir," replied the bombadier.

"Fine," came the pilot's reply. "You pray. Everyone else get ready to jump—we're one parachute short!"

"The shortage could have been serious," Darby admits, "but there just happened to be an extra chute on board. Sometimes on day flights we carried a photographer in the lead ship. He must have left his chute in the gunner's well, where it got covered up and was overlooked by the ground crew during their inspection. I believe it was Bennett who found it."

The smoldering fire flooded the bomber with smoke—an alarming in-flight situation. Then, as the men settled their nerves in anticipation of the bomb release and turn toward home base, without warning all four engines cut out!

T/Sgt. Frost, the engineer, was occupied in the top gun turret when the silence exploded around him. In the dim half-light of the late evening, he saw the four whirling propellers slow. Like everyone else on board, he froze momentarily as the bomber began to settle.

In the cockpit, the first stunned instant of shock passed as Berkeley and Murphy feverishly worked through the emergency procedures. Propeller pitch . . . mixture control . . . throttles . . . turbo-boost selector . . . and the four twelve-hundred-horsepower engines caught—and resumed their comforting drone. To a man, the crew went limp. A few even grinned. Then the engines cut out again.

"It got awfully quiet for a moment or two each time," Murphy recalls. "No one said a word, just worked."

During all the confusion, Darby was bent over the bombsight in the nose. He was holding the release button and waiting for the cross hairs to come on target. "In this type of business, these things are to be expected," he reflected. "I was busy and didn't give it much thought, but I wasn't exactly pleased with the situation, either. Despite the smoke, haze, and commotion, it was bombs away. The strike was one hundred percent effective."

Sergeant Salley, gunner in the waist position, remembered

those moments well. In 1965, while a member of Delta Air Lines Flight Control in Atlanta, he recalled, "I could only see the target now and then as there was considerable cloud cover. We were advised to expect heavy *ack-ack,* and possibly fighters. Rangoon was considered the toughest target in Burma but none of the Japanese antiaircraft bursts were close until our engines cut out, then we dropped directly into it."

As the B-24 slipped into the midst of the exploding shell bursts, it was rocked and buffeted by the shock waves. A few crewmen were certain whining shrapnel sliced through parts of the wings and fuselage. Because of the two power failures, the plane lost altitude it could never regain and it lagged still farther behind, unable to rejoin the other planes as they disappeared homeward over the Bay of Bengal. Again Berkeley and Murphy got the engines operating. As the minutes ticked by, the crew's frayed nerves slowly returned to normal. With an in-flight fire and two complete power failures behind them, they gained confidence with the thought that they had left their troubles behind, at least for this mission.

They were dead wrong; their headaches were just beginning.

William Frost, today Pan American Airway's Senior Maintenance Supervisor at Ascuncion, Paraguay, tells what happened next:

> Why all four generators failed is a mystery. We learned later that other B-24s had similar difficulty when their generator mountings broke, but it didn't seem possible that this could happen to all four of our generators. Shortly after we left the target, the first one went out, then the second one failed and nothing I could do would keep them on the circuit. We were still about an hour and a half from our base when the remaining two failed suddenly. I went to the APU [auxiliary power unit, an engine-driven generator used for emergency power] and started it, but it ran only a

few minutes before it blew the cylinder head. This left the storage batteries, which for some inexplicable reason, probably a short somewhere, drained dead in short order.

Scolavino, who also had duties as the assistant flight engineer, knew precisely how—and why—the APU failed. He recalls, "Frost sent me to start the APU; if it could be operated, we'd have enough electrical power for the aircraft. But because we were at altitude it wasn't a good idea to try to start it; fumes built up in the cylinder that could cause the cylinder head to crack or explode. We had no alternative, however. I tried again and again to crank it and, sure enough, when it finally started the cylinder head blew off. I picked it up by the spark plug and showed it to Frost. 'That's all she wrote,' he said."

Frost had other problems too. Serious ones. He was gravely concerned about his ability to control the bomber's propellers.

The propellers on a B-24 were controlled throughout their constant-speed ranges with an electrical step-head motor that was designed to hold the propellers at the rpm at which they were set if a complete electrical failure occurred. But they didn't hold; the blades kept creeping to a lower pitch little by little because of the air pressure against them. The engine rpms gradually got higher and higher, moving toward overspeed.

In the gathering gloom of the cockpit, Berkeley and Murphy were sweating out the instrument readings. As one generator after another failed, Murphy watched the tachometer needles inch toward the red danger line.

The engines were trying to overspeed all along. The hours crept by and, although the propeller pitch was supposed to be fixed, it slipped up a little at a time. Our propeller governors needed electrical power to reposition the control valves that held a constant engine speed, but now the slightest change in altitude caused our rpm to increase or decrease. We flew over four hours at night, trying to hold the same altitude. Once a propeller changed pitch, we had no means of bringing it back. When our troubles couldn't be corrected, we realized we were in deep trouble—night and over water—with more than four hours to go. I imagine each man did a little reviewing plus a prayer or two. I know I did.

As the battery energy drained to a trickle, the instrument lights in the cockpit grew dimmer until finally the interior of the throbbing bomber was in total darkness. The pilots peered out their respective cockpit windows at the faint outlines of the propeller arcs and watched the engine cowlings carefully. The first signs of overspeed showed themselves as the engines began to vibrate, slowly at first, then with a steadily increasing tempo. Every electrically operated instrument was dead now, and all communications were out. Even the intercom was useless. They had no lights for signaling, no way to send radio identification to friendly aircraft, or a distress call to search-and-rescue units.

Then Murphy remembered a personal item—his flashlight. He brought it out and trained it on the instrument panel. Aside from the engine gauges, only the vacuum instruments, the flight indicator, and the gyro compass were working. As long as the flashlight held out, they could navigate homeward. Of those apprehensive moments, Murphy said, "We dreaded to think of those batteries going dead, too. I took the flashlight with me when I abandoned the plane."

All the while, T/Sgt Frost and S/Sgt Scolavino, the assistant flight engineer, worked fratically to regain electrical power or even part of it. But their blind fumblings in the darkness of the bomber's belly were futile. The electrical system would never work again.

Now the blacked-out B-24 was moving over the coastal lowlands near Calcutta, still over two hundred miles from Allahabad. With engine operations touch and go all the way from the target back to India, Berkeley and Murphy were anxious to set their bundle of trouble down as quickly as possible—*if* possible. They banked the plane into a wide, cautious swing over blacked-out Calcutta and looked for an airfield. Dum Dum Field was down there somewhere . . .

That's when they discovered they were not alone in the sky.

Because their intercom was dead, no one knows who first spotted the new threat. It was probably Salley, who says of his frightening discovery, "I had been in the waist position for most of the mission, near my single fifty-caliber machine gun. It was just a little before nine. I was standing at the open left waist window and saw a plane make a pass from behind and above on my side of the ship. He made two passes that I saw, and each time I was frozen with the fear that he would open fire on us."

In the tail, S/Sgt Bennett also saw the plane zoom past, uncomfortably close. "We spotted another aircraft with a single exhaust, indicating it was a fighter type, presumably a British Hurricane and later verified as such. Besides an impending strafing, our engines were on the verge of disintegration."

As the British fighter plane whizzed over the Liberator and flashed past the cockpit windows, Berkeley and Murphy, with enough trouble on their hands to last them for the remainder of the war, were not heartened when they caught sight of the exhaust glow. Murphy recalls, "When the Hurricane made a pass at us in the darkness, we got ready to jump because we had absolutely no way to signal him—no running lights—nothing. The Japs had visited Calcutta before, and as we were heading directly toward it, he had every right to start shooting. But since we took no evasive action he let us go."

The big twin tail of the B-24! That's what the Hurricane pilot must have recognized on his last pass. After the second look, the fighter vanished into the night sky. Once more the crew breathed a sigh of relief. They had a little more time.

Their reprieve was short-lived, for now the serious condition of Number Three engine captured their attention. It was rapidly approaching the critical danger point. Frost knew what would happen when the overspeeding power plant reached its limit of endurance. "The rpms were gradually becoming excessive, and with no hope for electrical power, it was impossible to feather any of the propellers. I saw sparks and flames belch from Number Three. When it became very rough, Lieutenant Berkeley ordered me to notify the crew to prepare for bail out. I assembled the men, part on the bomb-bay catwalk and part in the cockpit."

Darby, in the nose, was unaware of the dangerous state of Number Three engine. "I had come out of the nose and was in the rear-gunner's compartment. Frost signaled me with a flashlight. I walked through the bomb bay to the engineer's station, and he told me the crew was preparing to bail out."

The decision to abandon the hopelessly failing bomber came none too soon. Murphy was nearest to the backfiring, and now wildly vibrating, engine and the danger was growing that flying parts would hurl from it at any moment and slice, shrapnel-like, through the cockpit. It had happened before.

"I was on the right side and watching the engine closely," the copilot reports, "so I had advance warning before it started to break up. As the propeller rpm moved close to three thousand, we knew it wouldn't last long even with the throttle cut back. With everyone in the bomb bay, we told them that when the doors opened (they were hydraulically operated), just go—don't ask questions. The propeller broke up first, then the engine wrenched itself completely out of the wing. We went then."

Craigie, Bennett, Darby, Frost, and Salley were the first to leave. Bennett remembers, "Frost told us to go as soon as the doors opened. Fortunately for Darby, the extra chute allowed him to join us on our trip down. He can talk about luck. Craigie left the ship seconds before I did."

With a forty-five strapped around his waist, Darby regarded the black pit only fleetingly before he dropped through the right side of the forward bomb bay. "After I opened my chute I reached in my coverall pockets to see if

I'd lost my cigarettes. I hadn't, and since I had nothing else to do, I lit up. It seemed the natural thing."

Frost took time to consider protection for himself and the crew once they had assembled on the ground. He was well armed, at least when he jumped. "We bailed out about nine-ten," he narrated, "and I went out the bottom with a Thompson submachine gun, 125 rounds of ammo and my musette bag. I counted to ten and jerked the ripcord. As the chute popped open, the loosely adjusted chest strap snapped into my face, cut my mouth and smashed my nose. Nothing serious, but when I came to my senses I was floating at five thousand feet with nothing left by my flight cap in my hand—and it was on my head when I bailed out. A few moments later I saw our plane explode and burn below us."

Edward Salley remembered his big step into dark space: "About ten minutes after the Hurricane scare, I got word by way of mouth (shouted from the catwalk above the engine noise) that we were to abandon the plane. When the doors opened, I dived out head first, clutching the ripcord in my right hand. I almost opened my chute too soon, because when it popped I saw the tail of the bomber right above me. The chest buckle hit me in the mouth and force my teeth through my bottom lip. I was only out momentarily and came to in time to see the plane hit."

For Adolph Scolavino, the last minute of the plane's flight under the pilot's control was jammed with urgency and suspense. "I was the sixth man out . . . left the plane at seventy-eight hundred feet. Then came Lieutenants Thompson, Murphy, and Berkeley. When I left the plane, Berkeley was still holding it from diving so we could bail out. I dropped through the rear bomb-bay door and counted to ten as I'd been trained. (It may have been a fast count though.) After my parachute opened and I got oriented, I saw the plane flash by above me, then disappear into the night, its right wing low and losing altitude rapidly. I didn't see it hit, but when the sky was suddenly lit up by the explosion and intense light, I mistakenly thought it was a Jap searchlight trying to pick us up as we floated down.

"I had no feeling of descending; I felt no wind coming up at me. Because it was night and I was a young twenty-two, all of 115 pounds in five feet, four inches, I was a light load for the chute; but I started to imagine all sorts of things. The first was: *'I'm stuck up here!'* The next: *'I'll bet I have a large twenty-eight-foot chute instead of the twenty-four-foot size!'* So I started pulling my riser lines to dump air. This caused me to swing widely back and forth, so I stopped pulling. Actually, I was descending at the proper rate, but in the blackness, without a reference point, I didn't know this. At one point I thought I was going to land in the center of a big city because the stars, reflecting in the water of the rice paddies below, looked like the lights of a large town."

Satisfied that the enlisted crew had cleared the bomb bay, the plane's officers prepared to exit. Murphy recalls, "Berkeley's order when the time came to bail out was a simple one—'Go!'

"I went out the bomb bay. Thompson preceded me, rather reluctantly as I recall. He said, 'Don't rush me . . . I'm going!' I was in a hurry as was Berkeley, who was the last one out. No moonlight—pitch dark."

The night stillness was broken only by the wind whistling through the riser lines to flap against the nine parachute canopies floating over the Indian jungle. Although each man bailed out only seconds apart, they had opened their chutes at different altitudes. This, and the wind, drifted them apart. In the moonless descent not one crewman saw another. Craigie suffered the most severe injury from the loose parachute chest buckles and he was bleeding profusely from a broken nose cartilage when he splashed into a lake. For nine hours, half swimming and half wading in rice paddies and jungle swamps, he fumbled about in the darkness.

Bennett landed uneventfully on dry land and immediately shed his parachute harness.

I called to Craigie, thinking he'd be close by, but I couldn't get him to answer. When I tried to walk around, I found I was practically surrounded by water, so there was nothing to do but try to settle down for the night. The seat cushion of my parachute made a good pillow and I lay down and stared up at the stars—until I became sharply aware of the weird noises coming from the rice paddies and swamps. My imagination had a field day. I'll never forget that night at the edge of the jungle, surrounded by a host of invisible enemies.

Finally, I fell asleep from sheer exhaustion and when I awakened it was daylight. I saw a native coming through the swamp in a dugout and called to him in broken Hindustani. I motioned for him to come closer, but it was of no use. After what seemed like hours, another native approached and I used every means of persuasion to convince him I was an American in need of directions and transportation. The way that finally worked was by waving money in the air.

The native took me to a small village where I quickly improved relations by feeding the small children chocolates from my survival kit. A boy of about fourteen told me—in broken English—there was a British airfield about twenty-five miles away. There was no transportation available, so we hunted around for something. The only thing that looked promising was an old bicycle with both tires flat. There was no pump to be had anywhere. Then I remembered the CO_2 cylinder (for inflating our Mae West life jackets) and I used it to inflate the tires. They held, and away we went, double-deck.

After a few miles my "faithful native guide" pointed out a railroad track and we followed it to a small telegraph house. The telegraph operator graciously served us tea and crumpets and stopped the next train to Calcutta. When I boarded it, I was surprised to find Salley and Craigie, much disheveled, but none the worse from the experience. We all started talking at once, excitedly asking about the other crew members. We had a few laughs, such as: Why did Craigie, our radio operator entrusted with carrying secret-code data, tear it into little pieces and let it float away on the water?

Salley and Craigie had met earlier near the crash site and boarded the train one station ahead of Bennett. Salley told of his frustrating inability to see the ground during his descent, and recalled:

When my feet hit a small tree limb, I knew I was near the ground. The limb broke under my weight and I landed rather gently. I recall that I sat for some time, shaken and scared, before I removed the parachute harness. I left it where it lay and began to look for the other crew members. I called out several times but no one answered.

I lay down in a plowed field with my Mae West for a pillow and went to sleep. Just before I dozed off, I remembered I'd left all my survival equipment in the chute back where I'd landed. My lip, though swollen and tender, didn't prevent me from sleeping—restless though I was.

Shortly after dawn I found a village. I was greeted with much suspicion until I made the natives understand I'd bailed out of an airplane. They communicated to me that some people from the village had already found my chute and brought it there. When I found someone who spoke English, I explained my plight more fully and as soon as he translated to the villagers they couldn't do enough for me. They produced the chute and I unzipped the back of the pack and took out chocolates, a machete, and the other survival items stowed there. I cut the pilot chute off (I still have it) and gave the main canopy to the natives. My sore lip prevented my eating anything solid, but I managed to take some of the chocolate and drink some coconut milk.

After a short rest, my English-speaking friend and I walked about five miles to another village on the river bank, where I met Craigie walking into the village followed by an Indian boatman carrying his chute. He was still dripping wet, tired from sloshing through rice paddies and swamps, and weak from loss of blood. His nose was very swollen and he had difficulty talking. Ironically, he was the only one of our crew who had heard the shouts of the others. He distinctly heard Bennett soon after he landed, but couldn't call back because of his injury.

Natives guided us to a railroad five miles upriver. The station was little more than a shack and very near our crashed plane. We walked over to it. It was scattered over a wide area and had burned on impact. We found a rather puzzling thing at the site: The bulletproof glass from the turret lay off to one side with hardly a scratch, while the turret itself was a twisted and crumpled mass of metal. We boarded the Calcutta-bound train and at the next stop Bennett joined us.

Darby, during his descent, scarcely had time to enjoy a few drags on his cigarette before he found himself on the ground.

I landed in a rice paddy on the edge of a jungle, and walked around until I reached a small village. I spent a couple of hours on the porch of a native hut until dawn. As I was preparing to move out, a native constable told me there was another American nearby. He pointed out the direction and we struck out with another native trailing behind with my chute. They took me to a British outpost on a river—and here I met Frost.

Frost often wonders who found his Thompson submachine gun (or if it was ever found), the 125 rounds of ammunition, and his musette bag. Without protection in the night, he wisely decided to play it safe.

I had no injuries on landing, and decided not to attract attention by calling to the others. I didn't know where I was or who might be around me, so I spread my parachute on the ground, pulled it over my head for protection against mosquitoes—and spent a very restless night.

Just after daybreak some Indian villagers came by, but

they were so frightened when they saw me they wouldn't talk. About mid-morning I met Darby along the bank of a canal and we got on one of the barges that took us to the outskirts of Calcutta.

Scolavino landed knee-deep in water in a rice paddy with a cut lip and thigh.

I crawled out and found a dry spot of land (it was pitch black), curled up in my chute, and fell asleep. About three A.M., I sensed someone nearby and opened my eyes. There, looking down on me was a native dressed in white cloth. I sat up quickly and reached for my knife, having decided not to wait to find out if he was friend or foe. When he saw the machete he ran.

I got up immediately and began to grope about; finally got to a river about daybreak. Two hours later I spotted a boat coming downstream paddled by two natives. I called to them, and when they rowed near, I took my book on Hindustani from my jungle kit and managed to make known who I was. They rowed me to a village where the headman gave me water and took me to a railroad station. I boarded the next train for Calcutta and the conductor showed me to a compartment where I found Lieutenant Thompson.

For Lieutenant Tom Murphy, a pattern of close calls was becoming evident in his wartime flying. On an earlier strafing mission over Akab, Burma, his B-25 was severely shot up by Japanese ground fire. The left vertical stabilizer was completely blasted away; one man was killed in the nose and two others were wounded. Then both engines of the crippled bomber quit twenty-five miles at sea. He and his pilot managed to ditch the struggling Mitchell safely and survive five days afloat in the Bay of Bengal by lashing together their five-man and two-man rubber life rafts. They had water but no food. About midnight of the fifth night, rough seas tossed them out of the rafts and two hours later they were washed ashore—back in Burma. They ran into a band of Ghurkas led by a British lieutenant. A radio message relayed through to Dum Dum Field brought a C-47 to their rescue. Later in the day it landed on the beach and picked them up—six miles above the Japanese lines.

So Murphy was an experienced hand at survival when he tumbled from the bottom of the condemned B-24. Like others of the crew, his chute chest strap slipped upward when the canopy billowed, and aside from a cut chin and lip, he suffered a large knot on his forehead. He kept his D ring as a souvenir.

> Except for a few strap burns and some other bruises, these were the extent of my injuries. On the way down I tried to stop the chute from swinging by pulling on the risers. But being a rank amateur, I dumped out all the air and fell—I don't know how far—before the canopy filled again. Needless to say I quit that.
>
> I made a soft landing in a rice paddy and didn't even fall down. Then I walked exactly fifty feet out of the paddy onto an earthen dam. I got plenty scared when Indian dike watchers found me, and a whole gang of them gathered nearby. Just a week earlier a British Blenheim crew was murdered by villagers twenty-five miles west of Calcutta when they made a crash landing. All I had to defend myself with was the machete from my parachute emergency pack. I stayed awake all night long watching them, but even so they managed to steal my pocketbook.
>
> In the morning I made my way to a village on a river bank where, to my amazement, along came Berkeley in a dugout canoe. We located the nearest railroad and boarded the first train for the city. Except for Frost and Darby, who joined us later, we had a gala reunion that night to celebrate our good luck. At the Grand Eastern Hotel, where we stayed for eight days, we had the chef cut the biggest steaks in Calcutta and fix them Western style.

Salley told of the crew's arrival in the city.

> We were met by Mr. Harold Marshall, Chief Inspector of the Calcutta Police, who escorted our disheveled troop to a hospital. Merrill's Marauders had just come out of the Burma jungle and had taken up every inch of space. We got a cursory examination, some minor patching, and were dismissed.
> There wasn't any reimbursement for any part of the expenses we incurred. In fact, we had to buy clothes to wear while we were awaiting transportation back to base. They were semi-British uniforms bought in the hotel's shop. One night, in the dining room, a British officer tried to throw us out. It seems the place was off limits to British enlisted men. We thought we'd have some fun, so we led him on until the situation got pretty heated. Our officers heard the ruckus and came over in time to straighten him out.

Darby and Frost were reunited with the remainder of the crew two days later when Harold Marshall brought them to a Calcutta theater during intermission. On November sixth, a plane was available to take the men back to the 436th, but there were not enough chutes on board. They arrived in Allahabad by train, and in a few days were reoutfitted and back on combat status.

Lieutenants Berkeley, Murphy, and Thompson were given another crew; Frost, after forty-one missions, was assigned to supervise aircraft maintenance. The remaining five of the original crew joined the team of the *Rangoon Rambler,* another B-24 whose record fortunately failed to produce another episode as singular as that of the jinx flight.

Scolavino returned to the States after fifty-six missions. He ended the war instructing B-24 engineer-gunners in Wyoming. He later rose to chief master sergeant with an F-102 Fighter Interceptor Squadron in Germany and retired from the Air Force in 1969. "I guess we were just plain lucky or it wasn't our turn," he reminisces. "Some people break a leg roller skating, and here nine men parachute at night over a country wild with rivers, jungles, and cobras—and end up with slight cuts. Sounds too good to be true, but it is."

Salley flew forty-nine missions with the Tenth Air Force and was discharged in 1945. Darby ended the war with thirty-seven missions. He reenlisted, was commissioned, and retired as a first lieutenant in 1959. His philosophy remains as cool today as then, when he peered through his bombsight on that black, shrapnel-filled night over Rangoon. "I think we survived because we were all previous servicemen," he said. "All well trained in our fields, with many years of experience among us—the type of men who don't get 'shook' easily, if at all. We knew our jobs and trusted one another in theirs."

Bennett came back with over fifty missions under his belt. He lives today in Charleston, South Carolina.

Only two members of the original crew are not living: John Craigie died shortly after the war and Ed Salley died in 1979.

All three of the bomber's officers remained in the Air Force. Francis Thompson became a full colonel, and following his return from the Far East, he remained in the service and attended pilots' school. Now retired from service, he makes his home in Dallas. William Berkeley retired as a lieutenant colonel in 1960 and became information officer for MATS at Scott Air Force Base.

Colonel Tom Murphy, whose oversize frame once snugly filled the cockpit of a B-52 or B-58, has flown Mach 2.1 in the Hustler. Now retired in Shreveport, Louisiana, after thousands of hours in military aircraft ranging from the B-25 to the B-58, his comment thirty-eight years after his narrow escape from death should make the heart of any patriot beat proudly.

"In the majority, it takes young men to fight wars, along with some mature leadership. I have no overpowering philosophy about these things. Luck? Yes—but the greatest gambit is the will to survive and go on again.

"Americans keep that with them always."

5

Ghost in the Desert

No one willingly travels the Sahara's "Desert of Thirst." No one, that is, except scientists who sound the barren wastes for traces of oil.

In November of 1958, two flying geologists of the D'Arcy Exploration Company, operating out of Cufra Oasis in Libya, sighted a ghost of World War II. It was an abandoned Liberator bomber, a B-24. The huge four-engine plane had bellied onto the desert floor twenty-five miles southeast of a gaunt, solitary limestone pillar named Blockhouse Rock.

Remnants of wrecked planes were still common near the North African coast, where smashed trucks, burned-out tanks, planes, and scuttled or torpedoed ships lay offshore as reminders of the struggle fought there a little more than a decade earlier. But here, 426 miles southeast of Benghazi, it was odd indeed. The bomber's faded pink color, blending at times with the surrounding terrain, told the geologist fliers it had once been assigned to desert region.

They dropped to a low altitude and circled. There was no sign of life. After marking the location on their map, they returned to Cufra and told the other members of the oil exploration party of their find.

It was not until the following March that the party's convoy of desert vehicles reached the sun-baked plateau, an unchartered land that simmered under a cloudless sky. Here in the dry dust where daytime temperatures reached 130 degrees and plunged to near-freezing at night, not a blade of grass could survive. Day by day, as the oil men pushed their

GHOST IN THE DESERT

U.S. Air Force Museum

The *Lady Be Good* rests on the floor of the Libyan Desert in the same position in which it came to a stop on April 5, 1943.

survey operations farther into the area marked on the map, they kept a sharper lookout. Then they saw it. Glistening in the heat waves that shimmered off the desert floor lay the long-lost derelict.

The bomber's fuselage had broken behind the wing when the plane skidded to a stop. The right outboard engine had twisted off on impact. Apparently, its propeller was still turning when the plane had pancaked into the sand years earlier. On the bomber's nose was a clearly discernible "64." The geologists jotted down other identification numbers, and on making a closer examination of the wreckage, marveled at how well the arid desert had preserved it. Water flasks and jugs of coffee were still full and palatable. Log books and instruments were unaffected. Gun belts were full; flight clothing hung on hooks. Cigarettes, chewing gum, and emergency rations were neatly stored away, and in the shadow of the plane's wings they found dried, mummified bodies of birds that had sought relief from the sun. The men fanned out to look for clues. They knew that, unless the crew had been rescued soon after they abandoned the plane, they had perished.

One of the men suggested that wandering nomads may have captured the crewmen and sold them into slavery deep in the Sahara. Perhaps, somewhere, they were still in custody. A Libyan guide shook his head no. This region, he said, was so uninhabitable that even desert tribesmen refused to enter it. "They say it has been cursed by God. It is called the 'Desert of Thirst.' "

When the men reached civilization in April, they notified Wheelus Field at Tripoli. A full-scale investigation sparked by the 17th Air Force and headed by Major General Spicer got under way. Responsibility was given the Army, however, because during World War II the Air Force was part of the Army. From the retired-records section at St. Louis came the first clue: 124301 had been missing since April 4, 1943, exactly sixteen years, with its nine American crewmen. It required two years of research and sixty pounds of records accumulated from four locations in the United States to tell the complete story.

The following month a two-man team from the Army Mortuary Service Headquarters in Frankfurt, Germany, flew over the bomber. By now, the records began to produce more information. The crew, as well as their mission, was positively identified; for the third time the Army Adjutant General's Office contacted the next of kin listed on the records. The first time had been shortly after the crew's disappearance, when they were declared missing in action. Then, a year later, another report was made when the men became legally dead under the Missing Persons Act.

The bomber was the *Lady Be Good,* named for Gershwin's popular song. It was once assigned to the North African Ninth Bomber Command, which was composed of the 98th and 376th Groups. They were known as the Liberandos. Their mission: to bomb Sicily and Italy, cut supply routes, knock down enemy air strength. They were based at Soluch,

U.S. Air Force Museum

The crew of the ill-fated *Lady Be Good*. From the left: 1st Lt. W.J. Hatton, pilot; 2nd Lt. R.F. Toner, co-pilot; 2nd Lt. Dp Hays, navigator; 2nd Lt. J.S. Woravka, bombardier; T/Sgt H.J. Ripslinger, engineer; T/Sgt R.E. LaMotte, radio operator; S/Sgt G.E. Shelley, gunner; S/Sgt V.L. Moore, gunner; and S/Sgt S.E. Adams, gunner.

one of the airstrips in the Benghazi area, on a base that was little more than a gritty, red-brown scar, scraped level through hard desert sand, and blown by the Libyan *ghiblis*, fierce desert sandstorms. It was a hellhole for the civilized soldiers who endured rationed water, dysentery, yellow jaundice, diarrhea, and eye infections. They ate their gritty rations standing up at mess.

The crew were old men by World War II standards. At the time when most crews consisted of nineteen- and twenty-year-old youths, the *Lady Be Good* didn't have one man under twenty-one.

First Lieutenant William J. Hatton was the pilot. He was twenty-six and he had left a bride when he departed overseas.

The other officers were Second Lieutenants Robert F. Toner, the copilot, who was twenty-seven; Dp Hays, the navigator, twenty-three; and John S. Woravka, bombardier, twenty-six.

Technical Sergeant Harold S. Ripslinger, the flight engineer, was twenty-two. The remaining enlisted men were staff sergeants; Robert E. LaMotte, twenty-two, was the radio-operator gunner. Vernon L. Moore, twenty-one; Guy E. Shelley, twenty-six; and Samuel E. Adams were gunners. All had met late in 1942 at Topeka, Kansas, for crew training and check-out in a B-24D.

Once assigned a bomber, they flew the South Atlantic ferry chain, and on March 27, 1943, they became part of the 514th Squadron at Soluch, Libya.

It was a dark beginning for Hatton and his crew, for now began a series of seemingly jinxed occurrences, bad luck, misfortune—call it what you will—that ended seventeen days later, four hundred miles in the desert.

As news developed about the finding of the missing bomber, rumors began to fly. One was that in 1943 a nomad camel caravan had seen an Italian armored convoy in the vicinity of the crash site. They had captured eight or nine Americans. Five perished and were buried by the Italians. But investigation ruled this out. Montgomery's Eighth Army had swept the desert clear of Italian and German forces by January of 1943—three months before the *Lady Be Good* crashed.

So, with no trace of the missing crew, the ghost bomber took on a mantle of suspense and mystery.

At Wheelus, plans were laid for a detailed air and ground search. A light Army L-19 preceded an Air Force C-47 to the scene. The party included the two mortuary men, Captain Fuller and Wesley Neep, and Captain J. M. Paule, an Air Force flight surgeon and expert on desert survival. Seven months had elapsed since the oil-company pilots had found the bomber. Everything was as their later ground party described it. It had remained in an unbelievably good state of preservation.

GHOST IN THE DESERT

The military men probed into the bomber's fuselage and noted two important items missing: life preservers and the hand-cranked Gibson Girl emergency radio. The plane's controls, they also noted, were not on autopilot.

The search party made camp and stayed two days. They took pictures, made sketches, and inspected the area for clues. During this time, the long-range radio in their C-47 failed. They discovered that the bomber had the same model. Curious, they removed it and installed it in the cargo plane. It worked. Radio failure on the *Lady Be Good* was not the reason it had strayed. The plane's two magnetic compasses and the radio-compass automatic direction finder also operated perfectly.

The missing parachutes clearly showed that the crewmen had bailed out. But where? Twenty miles away? Thirty? A hundred?

The bomber's nose pointed east. Were they flying in this direction when they abandoned it? Or were they flying south? Low flights in all directions around the plane revealed nothing. When they returned to Wheelus, the search party brought back enough questions to keep the world guessing as to the fate of the missing crew.

The mortuary men decided the only way to get the answers was to organize a full-scale ground sweep. C-47s airlifted supplies and men, Libyan natives were enlisted, and special ground vehicles were assembled for the largest search ever attempted in the Sahara. The Army furnished light aircraft, and the expedition, under Captain Fuller, pushed off in July. Working on the assumption that the men had bailed out north of the site, the searchers fanned out on a wide front and moved in that direction. At eight miles they found heavy tracks of Italian military vehicles heading north-northwest. They reasoned the tracks must have been there when the plane crashed. Had the *Lady*'s men found them, too? Two miles farther along the trail they had the answer.

"Over here!" one of the searchers called. They found a pair of aircrew high-altitude boots weighed down with stones. The toes pointed northward in a V. Their elation faded as they wondered how many of the crew this lone pair of boots represented. They still had no way of knowing whether or not the crew assembled after it bailed out.

They pressed on. The scorching afternoon sun grew hotter. The temperature climbed to 130 degrees. Here, a man's skin dries hard, sand dust penetrates the hair, clothing, and eyes. Surely the men of the *Lady Be Good* could not have survived long with their limited water supply.

The following morning they found the first of eight parachute markers—strips of cloth arranged in another V pointing along the trail. Someone had remembered their survival training. A few more miles along the trail six faded Mae West life preservers were discovered. All had their CO_2 inflation cartridges punctured. This told the searchers the crew had bailed out at night and that they had thought they were over water. Now parachute markers were found at regular intervals. Captain Paule was certain they would discover the bodies soon. He remarked, "Without water, they couldn't survive more than a day. With all the water they could carry, two days would be the absolute limit. They weren't riding either, they were walking." The trek was a veritable walk in hell all right, and no one in the party cared to dispute the captain's opinion. But the officer's estimate of the missing crew's endurance later proved to be in error.

Twenty-eight miles up the trail the motorized convoy found British tracks crossing the Italian ones. They came from the north-northeast and were identified as part of a huge British force that had joined the Free French forces in 1942 to destroy the Italian garrison at El Gezira, 140 miles southwest. Now the search party was faced with deciding which trail the crewmen had taken. They pitched camp for the night, and as they did, one of the men found another parachute marker a quarter of a mile up the Italian trail. Their question was answered for the moment at least.

The next day, the fifth, sixth, and seventh parachute markers were found. Then nothing. This had to be it, the searchers reasoned. The land vehicles spread out in a circular search pattern. One truck followed the Italian trail for another twenty miles until it disappeared in the dreaded Sand Sea at Calanscio, a region of shifting sand dunes that rise as high as seven hundred feet.

Captain Fuller decided to move the base camp forty-five miles due north of the *Lady Be Good*, between the Italian and British trails. He intended to search the entire northern part of the plateau. Then, one of the drivers found an eighth marker beside the British trail, pointing north-northeast. To the searchers, this could only mean the crew had split up after all, and would be found in two places.

The desert vehicles were equipped with sun compasses mounted on their hoods. This is the only reliable way to remain oriented in a land devoid of landmarks. As one truck held a straight course, another would follow at a faster speed, zigzagging across its path. Over one thousand square miles were covered and not one other scrap of evidence was found. Doubt began to creep into the men's tired minds. They began to suspect the bodies would never be found in the northern part of the plateau.

Thus far, all the search efforts resulted only in uncovering a preview of what must have been the fate of the missing men. An Arab nomad and his five camels were found where they had perished years earlier. The bodies were mummified, skin stretched taut over their skeletons. When they fell, the moisture in their bodies evaporated quickly, and with bacteria unable to survive in the arid, sterile desert, the remains petrified. The one-man caravan lay as it fell, undisturbed by the desert winds. Equipment found nearby indicated the man and his beasts died before 1900.

At this stage of the operation, General Spicer took a hand in the search. He flew to the base camp in late July and was briefed on the progress. He inspected the markers, checked maps, and conferred in detail with Fuller and Alexander Karadzic, the desert expert and its contractor of the vehicles. General Spicer decided to push for an all-out effort. Headquarters sent a C-130 to airlift Army helicopters to the site. It arrived with its cargo in mid-August. The whirlybirds penetrated twenty miles into the Sand Sea but found nothing. Next, RB-66s, jet reconnaissance planes, took strip photographs of the plateau and fringes of the Sand Sea. Nothing new was discovered.

Tired and disappointed, Fuller and Karadzic finally called off the search. The desert refused to give up the men after a demanding and expensive three-month search that covered over fifty-five hundred square miles. In approving the decision, General Spicer said, "Too many factors remain unknown—now perhaps forever—to make a definite conclusion." It was certain that the missing crew lay somewhere in the great Sand Sea and that only by an improbable coincidence would they ever be found. Before the search party withdrew, they returned to the *Lady Be Good* and painted markings on her to brand the plane as a derelict of the desert—a desert that had added a few more bodies to the thousands it already claimed.

Silence settled once again over one of the most puzzling air mysteries of World War II since attention had become focused on the *Lady Be Good*'s last flight and how it overshot its home base by 426 miles.

In March 1943, Lieutenant Hatton and his crew had ferried a brand-new B-24D to Libya, but when they landed at Soluch, the 514th was short of planes. Their bomber was turned over to combat veterans and for nine days they were a spare crew without a ship. On April 4, things changed. The crew of another B-24, the *Lady Be Good,* was in Malta with engine trouble on a spare plane they had been forced to fly temporarily. They left their plane behind for the ground crews to check over; now, with inspections completed, the *Lady Be Good* was ready to go. It was assigned on a temporary basis to Hatton's crew for a twenty-five-plane unescorted high-altitude mission that afternoon. The target was to be the marshaling yard and harbor at Naples. Two sections were to hit the target at sunset, scatter, and return home singly to elude enemy fighters. One hundred and six B-17s from Algeria were scheduled to strike the port city earlier in the afternoon.

The bombers were divided into two sections. Section A had twelve planes; Section B, which included the *Lady Be Good,* had thirteen. When Section A took off, they blew so much sand and dust behind them that the waiting planes of Section B took some of it into their air filters and intakes. The result was felt en route as Section B lost plane after plane due to engine failure, until finally, at seven forty-five, just five minutes out of Naples, only four planes remained in formation. The *Lady Be Good* moved up to lead position and following it were Lieutenants Worley, Swarner, and Gluck. Because of their delayed takeoff, these four trailed late all the way, and now, still several miles from Naples, the sun set. Section A had already blasted the target, but Hatton, knowing they would not be able to see the target in the darkness, canceled out. He led the remaining planes south, where they scattered according to instructions, dropped their bombs, and headed back for Soluch.

Then, except for three seemingly unrelated and insignificant reports, the *Lady Be Good* flew into history and was not seen again for sixteen years.

The next morning the *Lady Be Good* was the only one of the twenty-five planes unaccounted for. Lieutenant Worley remarked that Hatton's "64" had been with them until they turned back over Sorrento at sunset. The other pilots of Section B agreed. No doubt the plane had ditched in the Mediterranean, out of fuel. Then, one pilot, Lieutenant Ralph Grace, mentioned that he heard a B-24 pass directly over Soluch, headed southeast, sometime *before* midnight and *after* all the planes had returned from the raid. No one paid much attention to his comment.

They should have, for another pilot recalled that Hatton broke radio silence to get an emergency bearing from nearby Benina RDF Station. Had they checked the time of the call, simple logic would have told them where to send the search

The Lady Be Good *mistakenly received an "inbound" bearing after it passed over its home base at Soluch.*

planes that needlessly combed the Mediterranean for the missing bomber.

The tragic fact was this: Hatton had called for an *inbound* bearing *after*—not before—Lieutenant Grace had heard the B-24 pass over the airfield. It was at twelve-twelve to be exact. The Benina operator, assuming that the bomber must be somewhere north and heading toward them, turned his loop accordingly. He got an aural null as Hatton counted over the radio and replied to the pilot, "Your bearing three-three-zero magnetic from the station. Repeat. Bearing three-three-zero. Over and out." Actually, the *Lady Be Good* had already passed the station and was reading on the *back* side of the Benina loop. South. Its bearing was actually reading in the opposite direction, 150 degrees, and by now the plane was well into the desert. If Sergeant LaMotte had simply tuned in their Automatic Directional Finder on Benina, the ambiguity error would have immediately become apparent and left no doubt of the *Lady Be Good's* position with respect to the station. But he did not, and with the entire North African coast blacked out, the only way for the crew to spot the coastline on that moonless night was by the thin, light line of breakers against the beach. Somehow they all missed it. With no way of knowing the wind direction and velocity, and no sign of land beneath them, they had no choice but to hold course.

By one o'clock the *Lady Be Good* was 225 miles into the Libyan desert. A few of the crewmen grew suspicious that something was wrong. Hatton dropped lower. He knew they had had a tail wind up to Naples and would probably have a head wind back, but it was taking too long to reach the coast. He peered downward for a glimpse of something recognizable, but there was nothing. Benina had said he was 330 outbound, and he was making good the reciprocal bearing so they would not have anything to worry about. They droned on.

The fuel ran dangerously low after another forty-five minutes. LaMotte made an emergency call to Benina. No reply. A few minutes later the nine men stepped from the plane into the inky night sky. Two of them, Toner and Ripslinger, carried small pocket diaries. For almost seven minutes the Liberator lumbered onward, then all but one of its engines sputtered into silence. The bomber glided drunkenly downward, pancaked into the Libyan wasteland, careened sideways, snapped its fuselage in half and slid to a stop.

Not a living thing heard the sound.

On February 11, 1960, six months after the Army-Air Force team called off its search, a desert-supply pilot flew provisions to a campsite of the British Petroleum Company Limited, successor to the 1959 D'Arcy Exploration Company. He landed at Failing Cap, seventy-five miles north-northwest of the desert derelict. Before he took off for the return flight to Tripoli, the man in charge asked him to relay a message to Wheelus Field. His men had found five bodies of the bomber crew in the same area searched by the military team the previous summer and fall. A C-47 arrived the next day and the oil men led the party, which included a chaplain, to the final encampment of Hayes, Hatton, LaMotte, Adams, and Toner. They were lying close together amid their useless equipment—parachute cloth, jackets, shoes, canteens, and flashlights. Their silk "escape maps" were found, which, had they included another 120 miles south, might have shown the men the way to Blockhouse Rock and the nearest oasis at El Gezira. A sunglasses case with "Dp Hayes" imprinted on it left no doubt as to one man's identity. They had walked sixty-five miles in eight scorching days with little food and almost no water, to die.

The Air Force men reverently covered each body with an American flag, and the small group stood bareheaded in the sun as the chaplain offered a prayer for the spirits of the long-dead fliers and for their yet-to-be-discovered comrades.

Toner's diary, found nearby, attested to the unbelievable week in the desert Hades and told of other members of the party who had pushed ahead. His brief entry for Friday, April ninth, reads:

> Shelley, Rip, Moore separate and try to go for help, rest of us all very weak, eyes bad. Not any travel, all want to die, still very little water, nites are about 35 degrees, good N wind, no shelter, 1 parachute left.

Huddled with the other four, Toner lived to make three more entries while the others struggled on, through the sand dunes, slipping and falling, knee-deep in the soft dust, inching their way painfully forward, hoping that just over the next dune would be water. The half a canteen of water which they bailed out with was now empty. Each man was rationed to a capful a day.

Again, a massive air-ground search got under way. The helicopters, C-130s and RB-66s were recalled for "Operation Climax," but failed to turn up anything. Meanwhile, the British Petroleum Company men worked doggedly on. As the military searchers were again about to call off operations, the geologists found Ripslinger and Shelley, an unbelievable twenty-one and twenty-seven miles from where they found the first five. Shelley had walked ninety agonizing miles from the bail-out point.

The final entry in Ripslinger's diary was marked Palm Sunday.

> Palm Sun. Still struggling to get out of the dunes and find water.

The stamina of the men was, without a doubt, far beyond the expectations of survival experts. Somehow they must have drawn upon some source of power to keep going as long as they did. Their endurance gave survival schools much to ponder.

Army and Air Force personnel now surmised that the mystery marker on the British trail must have been placed there by John Woravka. They were certain his body, as well as Moore's, would be found in the great Sand Sea.

After seventeen and a half years, the seven American airmen went home with military escort for a long-delayed military funeral. When the last flier was laid to rest, the file was closed again on the last flight of the *Lady Be Good*.

Stubbornly, the file refused to remain closed. In August 1959, the desert gave up another part of the puzzle. Another group of the British Petroleum Company, working near the bomber, solved one mystery and created another. They found Woravka's body twelve miles north-northeast, in full view, where it had lain those many years. It was dressed in flying suit, Mae West life jacket, and parachute harness. The parachute canopy appeared to have only partially opened. John Woravka died quickly and mercifully. His parachute either fouled, or he had delayed too long in pulling the D ring and death spared him the ordeal of his fellow crewmen. A canteen three-quarters full of water was found with him. It was still potable and free from bacteria.

A mile south of the body, Air Force men found a stack of parachute harnesses and burned-out signal flares. Here the *Lady's* men had assembled and tried to contact Woravka. It was now clear that the Liberator's bombadier was not the man who made the lone marker on the British trail.

Who, then, did? Sergeant Moore? Hardly. He would have had to backtrack the twenty-six miles he came; yet, as the only member of the crew not found, his fate is still an enigma.

In July of 1962 the Air Force officially declared the *Lady Be Good* "surplus to Air Force needs" and announced that what was left of the B-24 would be abandoned in the Libyan desert.

If there was ever cause to brand a plane as jinxed, the *Lady Be Good* was it. It left a heartbreaking and costly trail. Things were going well for the 376th before the bomber joined up. When the *Lady Be Good* came along, it aborted its first mission and ended up with an escort of only three of the thirteen planes it started with. Eight days later its crew was dead on the Sahara; eighteen days after that Lieutenant Swarner was killed and two months later Lieutenant Worley and his crew were missing in action. They were never found. The sole surviving plane commander among the four who turned south over Sorrento was Lieutenant Gluck.

Later, in August of 1943, the 376th led 175 planes in the Ploesti oil-field raid, the most disastrous single American air strike of World War II. The lead plane plunged into the Mediterranean after takeoff, killing the crew. Twelve more aborted for various mechanical reasons before they reached the target. Forty-one B-24s were lost over the target, thirteen more on the way home, and 440 airmen were killed or listed as missing.

The jinx of the *Lady Be Good* was far-reaching. Its radio receiver was installed in the first C-47 to land at the desert crash site. Less than a month later, this plane was caught in the grip of a *ghibli,* a fierce wind storm, and forced to ditch in the Mediterranean. One propeller sheared off on impact with the water, whirled into the cockpit, and killed the pilot.

Other planes that used the *Lady's* parts had unexpected difficulties. A C-54, in which several of the bomber's autosyn transmitters were installed, developed propeller trouble. It reached an airfield for a safe landing only by throwing its cargo overboard. A single-engine Army Otter followed the ill luck of the Air Force cargo plane. While at the crash site of the B-24 earlier, its crewmen salvaged the *Lady Be Good*'s armrests and installed them in their plane. Eight months later, this plane, with ten aboard, flew into another storm over the Gulf of Sidra and disappeared. No survivors were ever found, but later, scraps of wreckage drifted onto the Mediterranean shore. Among the debris was the *Lady*'s armrests.

Although the shell of the crumpled bomber still lies baking in the Libyan desert, from the assortment of parts that were returned to the United States for technical study, many found their way to several military exhibits. The United

States Army's Quartermaster Museum at Fort Lee, Virginia, displays fifteen artifacts, mainly crew clothing and personal flight equipment. Among them is a canteen, two watches, a flashlight, and a Mae West life vest. Among the bomber's parts on display in the Air Force Museum at Wright-Patterson Air Force Base in Ohio are the *Lady's* Number 2 engine, a propeller, an aileron, one of the fifty-caliber machine guns, several instruments, and the nosewheel. Two display cases hold maps and documents taken from the plane soon after it was discovered.

Mysteries of the *Lady Be Good* persist, of course. What happened to the *Lady's* emergency transmitter? Who placed the eighth marker—the one pointing north-northeast—on the British trail? And the most haunting question of all: Where is the body of Sergeant Moore?

In the final analysis, will petroleum company scientists provide the answer? The odds against improbable coincidence furnishing the answers diminish every day. Oil men, however, will continue to crisscross the desert wastes, and if Sergeant Moore perished as did the others, eventually his body will be found. Earlier wanderings of the oil men repeatedly uncovered what the combined and organized efforts of two military services could not.

Unfortunately, with the ever-changing world situation, there is no way to be certain that the sergeant's remains have not already been located. Colonel Dennis E. McClendon, USAF Retired, whose early research resulted in the definitive book *Lady Be Good* in 1961 said, eighteen years later:

> To my knowledge, Sergeant Moore's remains have never been found; however, his body was left behind the Oil/Moslem Curtain when Moammar Khadafy ran us out of Libya—so who can tell?

PART TWO
Trailblazers and Adventurers

6
Where Is Salomon Andrée?

Years before there were such things as aeroplanes, years before the Wrights made their first run, skip, and jump down the slopes of Kitty Hawk in their flimsy gliders, a perplexing air mystery was born—aviation's first. Three daring balloonists who battled the Hell Wind vanished into the Arctic night. The greater part of the mystery, from its beginning in July of 1897, lasted for thirty-three years and left behind the question that is still unanswered: How did they die?

Salomon Auguste Andrée was a Swedish engineer and scientific aeronaut. At forty-three, he dedicated his life to two things: the unexplored polar regions and ballooning. With ten years of Arctic experience, he was certain he knew its winds and ways. He staunchly believed that from Dansk Gatt in northwest Spitsbergen he could pilot a balloon expedition across the North Pole to a mainland on the other side. Such a fantastic proposal in the late 1800s tempted newspaper readers to picture the tall, distinguished man first as a visionary, then as one with a touch of the daredevil. But polar explorers and fellow scientists knew Andrée better; he was simply a determined man with a strong personality and a stable temperament.

In 1876, twenty years before his departure into the barren ice fields, Andrée came to America with three dollars to his name and a limited knowledge of English. The young man wanted to meet the world's foremost aeronaut, John Wise, whose writings he knew by heart. He wangled a porter's job at the Philadelphia World Exposition where the sixty-eight-year-old Wise was to make a tethered ascent. Shouldering through the crowd, Andrée managed to lend a hand in running the ropes out and hauling them in for the landing. When the exhibition was over, he introduced himself and begged Wise to give him balloon lessons. Wise looked the young engineer over carefully and was impressed with Andrée's sincerity. Yes, he would help; he offered his knowledge and the hospitality of his home.

The old man and the young man worked together for many weeks. Wise taught Andrée how to select balloon cloth, how to design an aerostat, apply varnish, and make a gas generator. He suspected Andrée had some great project, some burning ambition in mind, and he finally asked the Swede point-blank one day, "Why are you really so interested in balloon flying, Salomon? What do you plan to do with your knowledge?"

Andrée replied without hesitation, "I'm going to reach the North Pole in a balloon!"

Wise could scarcely believe his ears. Andrée continued. "Ships have failed, sleds have failed. The best of men—Nansen and Abruzzi—have failed. How else but by balloon can man reach there?"

"But the weather! The winds!" Wise protested. "No one knows them!"

"I will know them," Andrée replied confidently. "I'll study them. And when I come back, I will tell the world what's up there—whether there's land, or people, or just a great frozen sea."

Back in Sweden, Andrée kept working toward his goal. In 1882, the First International Polar Year, he joined Dr. Nils Erkholm in setting up bases on polar tracts. He collected weather data, made exhaustive notes on wind directions and velocities, and tried to learn as much as possible about the mysterious polar regions. But when he tried to raise money for aeronautical research (which meant flying a balloon in Arctic weather), the response was as cold as a polar midnight. Finally, in 1893, the Lais Hierto Scientific Foundation gave him funds to obtain a small balloon, which he named the *Svea*. He succeeded in flying it for a twenty-five-mile stretch and reached an altitude of nearly eleven thousand feet. Within a year, Andrée had crammed as much flying experience under his belt as most aeronauts gained in ten. Twice he was marooned in the Baltic Sea after being wafted over wasteland and open sea at the wind's will. Each time he was rescued by a passing fishing boat.

In 1895, Salomon Andrée spoke before the Swedish Academy of Science to present his proposal on reaching the North Pole by balloon. With confident authority he reeled off, one by one, precise facts concerning the venture. Specifically, he would need an aerostat of sixty-six hundred pounds lift with a gas volume of 212,000 cubic feet, three men, scientific instruments, a four months' supply of food, and equipment for safety. An average south wind of only sixteen miles an hour from Spitsbergen would carry them to the vicinity of the Pole in two days. With more favorable winds, another four days would put them on land off the Bering Strait. He went on to explain how the balloon he had developed would not have to be at the mercy of the winds. Andrée wanted more control over his balloon than earlier aeronauts dreamed possible. He had devised an ingenious system by attaching two adjustable sails that ran up the side of the gas bag, together with a series of long steel cables that descended from the gondola to drag across the ice fields. By adjusting the sails and the length of the drag lines he found he had limited steering control. The friction of the lines against the surface would cause enough relative motion between the wind and the balloon to allow him to control his direction of flight by as much as forty degrees. It would work, for he had already tested it successfully on flights with earlier balloons. He had tremendous confidence in his balloon. In such matters as polar balloon flights, Andrée believed one had to aim high and combine the maximum of audacity with the maximum of preparations and precautions. When the cables trailed along the ground they also lightened the balloon of their considerable weight. When they were pulled up, their weight would cause the balloon to descend. He had modified an ordinary balloon into an unpowered dirigible with directional control.

With his study of the far north, Andrée was certain the Arctic winds during July generally blew from the south to the north for a month. He thought the permanent presence of the Pole's summer sun above the horizon would maintain the lifting power of the balloon by heating the hydrogen gas in the bag. The temperature would determine its expansion, and he would keep a constant height of eight hundred feet by the use of the guide ropes.

To Andrée's complete amazement, the academy endorsed his plan. They decided it wasn't as wild an idea as they had imagined earlier. Now, Andrée worked furiously to raise money for the expedition. He lectured and wrote articles. It was not enough. Then, fortune smiled as Alfred Nobel, Baron Oscar Dickson, and King Oscar II of Sweden provided the thirty-six thousand dollars needed. Andrée was on his way. He asked balloon manufacturers for bids and began to stock equipment. The contract for the largest balloon the world had ever seen went to the French constructor Henri Lachambre, who made the great gas envelope out of pongee mixed with wool and floss silk.

Andrée was a rather tall man with a bushy, mahogany-colored moustache and heavy, handsome features. He was now head of the technical section of the Swedish Patents Institute. There was, personified in him, the rapidly growing belief that man was destined to conquer the air. Before anyone else, he understood the advantages balloon flying in the Arctic had over the torturous marches across the rough ice fields and treacherous water channels or "leads." He wanted to see the Arctic from the air in a gas-filled balloon, and he trusted in the winds to carry him there. But there was more to it than that; he wanted to attain the greatest prize the Arctic could offer—the honor of dropping the Swedish flag on the top of the world and sailing beyond, to Alaska.

There were parts of Andrée's life as unknown as the

territories he proposed to explore. And there were many contradictions in his character. He never married. Music did not interest him. He did not read for pleasure and distrusted those who wrote for a living although, ironically, he once won a prize for an essay defining the gaps in the education of modern young women.

Of the many volunteers for the polar flight, only the qualifications of twenty-four-year-old Nils Strindberg satisfied Andrée. He was an expert in photography and meteorology, a good cook, and had studied the latest discoveries in electricity. Young, athletic, and enthusiastic, he was filled with faith in Andrée's plan. As soon as he returned from the flight he intended to marry. Andrée also wanted Nils Erkholm with whom he had worked earlier, as the third member of the party, but the professor was hesitant. He was leery of balloons and their flimsy-looking rigging, but finally, after two months of indecision, he said yes.

The North Pole expedition ship *Virgo* left Sweden in June of 1896 and arrived at Danes Island three weeks later. A fresh snow had buried the building material for the balloon shed and caused a three-weeks' delay. Then a dense fog moved in. It was not until the end of July that the gas generator could be set up for discharge, and the balloon, named the *Eagle,* finally stood, inflated and ready, in its crude wooden shed. Sometime before August twentieth, the polar expedition would have to lift off, because then the *Virgo* would have to sail for home to await the passing of winter.

Andrée supervised the loading of the balloon basket. There were three sleds, a canvas boat, guns, ammunition, camera equipment, scientific gear, and food. There was a new Primus stove, designed for cooking, lighting, and melting snow for water, three dozen messenger pigeons, and thirteen wooden keg buoys with records of the expedition's progress to be dropped on the icy ocean.

Now, the trio scanned the skies impatiently. They needed the right wind. Several times a stiff south breeze picked up, and the men hurried to the balloon basket. Each time the wind died and reversed itself. Erkholm became increasingly nervous with each false start. Andrée, watching him, knew the professor had serious reservations about the flight.

As they waited, Andrée talked with Captain Sverdrup of the vessel *Fram,* docked at the island. Sverdrup inspected the balloon in its shed and wished Andrée the best of luck. "When will you start?" he asked.

"As soon as we receive a favorable wind," Andrée replied, "which, as you know, is already well past due."

"What do you consider as a favorable wind?"

"A steady wind from the south, of course—as is usual here during the summer months," Andrée replied.

"Why do you say the usual wind direction in summer is from the south?" Sverdrup asked.

"Because I have studied the observations of the Arctic explorers," Andrée said. "Nordenskjold, for one."

The captain disagreed. "I'm sorry," he said, "but our observations don't agree with those of Nordenskjold and the others. There is no prevailing wind from the south in these Arctic regions during the summer."

Andrée became silent. The impact of Sverdrup's words began to work on his confidence. The seaman had many years of experience in these waters. Could he be right?

Then, shortly afterward, another crushing blow fell. One day in early August, Erkholm summoned them to the balloon shed. Inside, he led them to the upper scaffolding. From over the aerostat's great dome the professor pulled aside a fabric strip that covered a stitched seam. They all heard it; hydrogen gas was hissing out. A leak. Andrée returned to his quarters on shipboard to mull over the depressing state of affairs. He decided there was little hope of making repairs and getting the right wind before the twentieth, so on the thirteenth he released eight of his pigeons with this message:

From Andrée's Polar Expedition
Dane's Island
August 13, 1896
No South Wind. Returning to Sweden.

ANDRÉE

Andrée identified his homecoming with personal failure, but underestimated the enthusiasm he had sparked in the polar venture. The Swedish government gave him more money and the use of a gunboat to take another expedition to the jumping-off place. Alfred Nobel offered to finance a new balloon, but Andrée politely refused. This one, he said, would be airworthy with repairs.

Nils Erkholm withdrew. Andree felt it was just as well; Erkholm was nervous and high-strung and had announced that he decided Andrée's balloon, as presently constructed, was not capable of reaching the Pole. His replacement was Knut Fraenkel, a young civil engineer who was more than eager to see the unexplored Arctic from a balloon gondola.

In May of 1897, the men were back on Danes Island with their work party. By June, their aerostat had been inflated and tested. Again it was loaded. And again the false winds came and died. On July eleventh, at three A.M., the officer of the watch aboard the *Svenskund* reported gusts of wind from the south. Andrée was awakened, arose immediately, and went to his observatory. A brisk, galelike blow picked up from the south and within a half-hour formed whitecaps in the bay. But would the winds die again, as they had done countless times before? The rawboned explorer waited.

At eight A.M. he said he would have to consider the conditions carefully. By nine o'clock the sky had cleared; clouds were scudding briskly northward at last, driven by a strong current of the south wind. He went into conference with Fraenkel and Strindberg. If the *Eagle* could be caught up in the air current it would carry them to the Pole—and beyond. A quick look at the barometer. It was rising. Yes—this south wind would last for days! Andrée watched the waters churn under the wind and suddenly he knew their moment was at hand. He left the gunboat and went ashore to order the balloon shed demolished. As all of the south wall fell, he walked around the *Eagle* for a final inspection. Strindberg hurriedly took pictures. He handed the exposed plates to the gunboat's captain along with a hastily scribbled letter to his fiancée. The ground crew cut loose several ballast bags from the basket and the huge bag rose a foot off the ice, kept in check now by only three taut and straining ground ropes. There was a moment of hushed silence. It was time. Andrée looked about him at the men who had shared the hopes and labors of his twenty-year-old dream. Then, impulsively, Andrée, Strindberg, and Fraenkel embraced the workmen. They all wept unashamedly. Andrée pulled away, hoisted himself to the basket, and called, "Strindberg! Fraenkel! Let's go!" As the two men clambered aboard Andrée called out to the ground crew, "One, two, three, *cut*," and three knives slashed down in unison to sever the restraining lines. The *Eagle* trembled for a moment, then moved heavily upward. Its fabric steering wings billowed out as it pointed northward across the bay. Shouts and cheers spurred the balloonists on, but halfway to the mountains on the other side, the *Eagle* narrowly averted disaster. It faltered and dipped toward the icy water. Observers at the camp saw ballast hurriedly cut free, but still the balloon fell. It touched the water, bounced back into the air, and struggled for lift. Slowly it rose again.

Alexis Machuron, a member of the base party, stood with others watching the aerostat grow smaller. He described the last sight of the *Eagle* this way:

> The balloon travels on, maintained at the same altitude by the guide ropes. In the neighborhood of the hills there is an upward current of air; the balloon will follow this . . .
> We see it clear the top of the hill, and stand out clearly for a few minutes against the blue sky, and then slowly disappear from our view behind the hill.
> Scattered along the shore, we stand motionless, with hearts full, and anxious eyes, gazing at the silent horizon.
> For one moment then, between two hills, we perceive a grey speck over the sea, very, very far away, and then it finally disappears.
> The way to the Pole is clear, no more obstacles to encounter; the sea, the ice field, and the Unknown!

We look at one another for a moment, stupified. Instinctively we draw together without saying a word. There is nothing, nothing whatever in the distance to tell us where our friends are; they are now shrouded in mystery.

As the *Eagle* drifted into the northern mists, there were excited shouts as the men left behind saw something that had previously escaped their notice. About two-thirds of the aerostat's guide ropes had been left behind on the ground. When the balloon was launched, its lifting power had not drawn all of the heavy guide lines up with it; some were ripped away entirely because the lower coils had been wedged. The greater part of the famous cordage with which Andrée reckoned to guide his balloon lay useless on the beach; his balloon was now practically impossible to control in any gust of wind, and the slightest mishap could wreck it. It was no longer an unpowered dirigible making its way northward, but a balloon practically out of control.

When word reached Europe that Andrée's expedition was aloft at last and headed for the Pole, the news was flashed to the northernmost telegraph outposts of every country and then, more slowly, to the mining, fishing, and trading posts. Eventually, word reached the fur country, the farthest populated reaches of Arctic Siberia, Canada, Greenland, and Alaska. As the world grew hungry for news, strange things began to happen in the northern reaches. After several days, vague reports claiming to be sightings of the polar balloon drifted to news services from all over the world's northern wastelands. In a sense it was expected, for Swedish authorities had made certain that whites and natives in the extreme outposts were informed of the balloon's departure and were cautioned to be on the lookout.

On July thirteenth, the bark *Ansgar* was plowing east, two days out of North Cape, Norway, and eight hundred miles south of Andrée's takeoff point, when its crew sighted a "downed balloon." When the ship reached Denmark in early August, the crew said they were certain it was An-

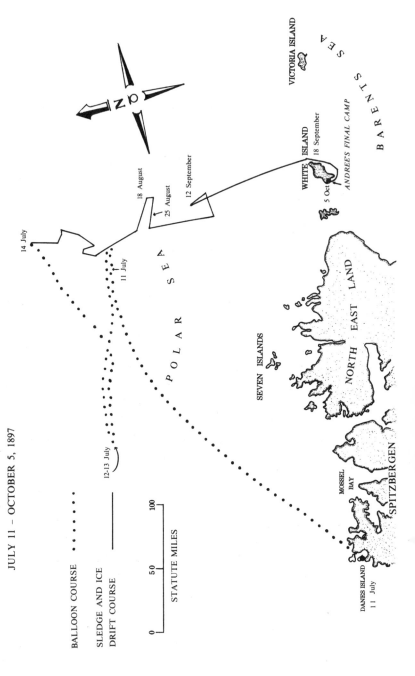

drée's balloon. They gave the details to the newspapers. It was black in color and "some of the gas had leaked out." They said it was "two fathoms above the waterline and covered with a net."

It was not Andrée's balloon; it was later proved to have been a dead whale.

On July seventeenth, a woman in a small Swedish town, "whose truthfulness was beyond question," had gone to her window to pull the shade, when she saw what she took to be a balloon with drag ropes and a net. She said one man was in the gondola.

A dispatch from Stavanger, Norway, dated August thirteenth, reported that the crew of the steamer *Kong Halfdan,* off Norway between Haugesund and Ryvarden, had sighted a "big balloon." It passed so close to the ship that "its drag ropes were seen."

The scientific world waited as the months slipped past, waited and wondered. Where was Salomon Andrée?

Searches began, sparked by an item that appeared in a Swedish newspaper in May of 1898.

ANDRÉE'S BALLOON FOUND
IN EAST SIBERIAN FOREST

Andrée's balloon has been reported found in a forest in eastern Siberia. The Arctic explorer with two companions ascended near Spitsbergen in July 1897, in an attempt to reach the Pole. It was the last seen of them. The Swedish government has ordered an investigation of the report.

It was possible that the *Eagle* could have drifted into central Arctic Siberia; so, in the summer of 1898, the J. Stadling expedition journeyed overland from Stockholm to the Venisei River, well north of the Arctic Circle. Fraenkel's brother, who was an eager member of the party, accompanied Stadling, but the expedition contributed nothing to the determination of Andrée's whereabouts. At the same time, Walter Wellman was leading an expedition to the Franz

Joseph Islands. He was convinced that Andrée could have made Cape Flora, where a store of provisions had been left by an earlier party. But Wellman found no sign of the missing men and abandoned all hope of finding them alive. In the February 1899 issue of *Century Magazine,* he wrote with an unmistakable air of finality, "Poor Andrée! Poor, brave, dead Andrée!"

A. G. Nathorst, the famed Swedish polar explorer, set to work to interest the Swedish government in a search of northeastern Greenland. In the spring of 1899, he was off, riding a wave of popular sentiment and government sympathy for the lost aeronauts. In the far northern regions he talked with travelers who had seen "strange pigeons" flying about (Andrée's messenger pigeons?), and one man reported that natives had killed and eaten a bird unfamiliar to them, and lost a message that had been fastened to it.

Between Northeast Land in Spitsbergen and Victoria Land lies White Island, which Nathorst twice visited during this time. From his ship, the *Antarctic,* he landed small parties on the north and south shores and expressed his theory that Andrée's balloon had "probably drifted close to this place." Then he became absorbed in studying the island's plants and moss. On the south shore his people walked about on a small track of flat land near the shore, where a new-fallen blanket of snow lay, unaware of what it covered.

In 1910, there was one last flurry of interest before Andrée's disappearance faded from the public mind. In that year, a Canadian newspaper carried the account of one William Irvine, a seventy-one-year-old trapper who reputedly was told by Eskimos in the Upper Hudson Bay that, two years earlier, Andrée's expedition had landed there. There was a gunfight and all had been killed by the natives. Irvine concluded by saying, "They'd be hard to find now."

All the stories seemed to have one thing in common: They could not be proved.

Well, then, where was Andrée's lost expedition? Surely, explorers pondered, there must be some valid record, some trace of their fate. But, except for the finding of two messenger buoys and a single pigeon, the world last saw Andrée on that gray, leaden day when he faded from view over the mists of Virgo Bay. Four days after the balloon's departure, the Norwegian sealer *Alken* was a hundred miles north of the takeoff point. A pigeon alighted on the rigging, carrying a message from Andree. It read:

> From Andrée's Polar Expedition to *Aftonbladet*, Stockholm.
>
> July 13
> 12:30 Midday. Latitude 82 degrees 2' Long. 15 degrees 5 E. Good speed E. All well on board. This is the third pigeon post.
>
> ANDRÉE

Of the thirty-six birds carried by the balloon, this was the only one ever accounted for. Of the thirteen message buoys, only two were found. One, which contained a brief note, was picked up on May 14, 1899, by two fishermen who found it floating in Kolla Fjord on the north coast of Iceland. It had been thrown out a few hours after takeoff, and had drifted for 672 days. The other buoy was found on August 27, 1900, where it had been forced ashore at Finmark, Norway. It had been dropped before the other and drifted for 1,142 days. Its message read:

> *From Andrée's Polar Expedition*
> July 11, 1897; 10:00 G.M.T.
> Buoy No. 4, the first thrown out.
> Our journey has hitherto gone well. We are moving on at a height of 830 feet in a direction which at first was N. 10 degrees E declination. Four carrier pigeons were sent off at 5H40 p.m. Greenwich time. They flew westerly. We are now in over the ice, which is much broken up in all directions. Weather magnificent. In the best of humors.
> ANDRÉE STRINDBERG FRAENKEL
>
> Above clouds since 7:45 G.M.T.

Thirty years passed after this message was found. Andrée's whereabouts now became part of Arctic legend. Seamen told and retold the tale in every ship's forecastle. Then, in the summer of 1930, the Andrée mystery was suddenly revived. It centered, strangely, on White Island, which in itself was always a rather mysterious place. Barren, windswept, and desolate, the small island in the Arctic Ocean is strewn with rocks and granite, its only vegetation nonflowering plants.

In July, the polar explorer Theodore Grödahl landed there. He was searching in the *Hanseat* for the remains of the airship *Italia* that had disappeared two years earlier. Except for a few old tins scattered in the snow, he found nothing of interest, and left.

A few weeks later, in August, the *Bratvaag,* a Norwegian sealer captained by Pedar Eliassen, plied its way through the floating icebergs that surrounded the elliptical-shaped island thirty-five miles east of Spitsbergen. For this voyage the ship's crew was host to geologist Dr. Gunnar Horn of the Norwegian Spitsbergen Polar Sea Research Institution. When walruses were sighted on shore, the ship anchored and landed two whaling boats. One crew went to hunt animals while Olaf Salen and Karl Tusvik looked for fresh water. A new snow had fallen, but here and there the rocky terrain showed through the white blanket. On a slope a hundred yards from the harbor, the two sailors found a small brook and followed it inland. After a short trek they paused to rest and drink. As one of the men bent over the clear water, the dull glint of a metallic object caught his eye. It was half-buried among the rocks on the other side of the stream. He pulled it loose and wiped it clean. It was an aluminum pot lid, a strange thing to find there.

The men looked around carefully. A few minutes later they found one end of a canvas boat protruding from under a snow drift near a cliff.

Dr. Horn and Captain Eliassen were on shore when the sailors returned, breathless, to tell of their find. As soon as Eliassen saw the aluminum lid he exclaimed, "Andrée!"

Back at the canvas boat the landing party scraped the ice and snow aside. There were oars, instruments, rifles, cooking pans, tinned food, boxes of ammunition. They uncovered a brass boat hook of the late 1800s on which was engraved: ANDRÉE'S POLAR EXPEDITION, 1896. There was no longer any doubt that thirty-three years ago, long before many of the *Bratvaag's* crew were born, Andrée, Strindberg, and Fraenkel had sought safety on White Island.

Was this merely part of their abandoned equipment? Had the explorers pushed on to the peaks of Northeast Land, which they must have seen in the distance? Captain Eliassen answered the question a short time later when he and Dr. Horn discovered the first body. It lay "leaning against the slightly sloping wall of rock" above the canvas boat. The skull and upper torso were missing, having been, presumably, carried off by bears. The skeleton's bare knees were seen through its tattered trousers; the lower ribs protruded through a rotted jersey jacket, inside of which was sewn the letter A. That this was the remains of Salomon Andrée was confirmed when both a large and a small diary were taken from the inside of the jacket. Although the record books were frozen solid, with careful handling they later told in Andrée's own words the details of the party's amazing adventure. After thirty-three years his icy phantom rose suddenly from the Arctic wastes. A voice from the grave now took up the long-stilled part of the story and for the first time the world learned the details of the strangest of all polar tragedies.

Lying near the bones was a rifle, more ammunition, eating utensils, and a package of photographic plates. Of the fifty-odd exposed negatives, twenty were developed successfully. Of these, twelve were reasonably clear photographs taken by Strindberg thirty-three years earlier. His camera had a self-timer and several photographs showed all three men together. All proved to be of the balloon flight and the sled journey; none were of White Island scenes. The Primus stove was found half-filled with fuel oil. Dr. Horn noticed the air valve was closed. He pumped up the apparatus, opened the valve, and lighted the fine spray. When tested, it could still boil a liter of water in six minutes. It worked perfectly. Perhaps, as Dr. Horn was to surmise later, it may have worked too well.

Beside Andrée's body was a smooth place free of rocks. Here were found most of the party's supplies, another sled, a sleeping bag, matches, utensils, medicines, and Fraenkel's pocketbook. This was the site of their tent.

A hundred feet to the north, a sailor discovered another body. It was wedged in a cleft between two large boulders, covered with a layer of small stones about a foot deep and arranged in a burial cairn as protection against predatory polar bears. Carefully, the men separated the bones from the rubble. On the remnants of the skeleton's jacket were the faded initials N. S., and a gold locket was found with the still-discernible photograph of a girl. This, then, was Nils Strindberg.

Within the hour the skies churned up a threatening storm. At this time of year the weather could change quickly, so Eliassen ordered the crew to gather their belongings and return to the ship. By the time they reached their beached boats, the swelling waves were pounding the shoreline. Soon afterward the *Bratvaag* set out for home and the news was radioed for relay around the world: *"Andrée's expedition found on White Island."*

Newspapermen thronged to Skjerno, a small village on Norway's north coast, where the *Bratvaag* would make port. Foremost in their minds was the question, "What happened to Fraenkel?" Had he drowned? Was he carried off by bears? Maybe he fell from the balloon. Andrée's diaries would surely give some clue, but it would be weeks before they would be ready for deciphering. Swedish journalist Knut Stubbendorf could not wait. Even before the *Bratvaag* landed, he set off in the sealer *Isabjorn* for Andrée's camp, in the very face of the coming winter. Fortunately, the elements smiled on him, for the snow had melted even more than when Horn and Eliassen were there and several things they had overlooked were now laid bare in the bright sunlight. He found a third sled, a human backbone, thighbone, and pelvis. Whose were they? Certainly not Andrée's or Strindberg's. Stubbendorf suspected that Fraenkel did not perish before the party reached White Island. He searched around the site carefully and found Strindberg's logbooks and almanac. There was more equipment and the remains of game they had killed.

The journalist paid close attention to the cleared space where the tent had been, and nearby he found something else the *Bratvaag's* crew overlooked in their haste—a human skull. It later proved to be Andrée's. He found fabric from the *Eagle* and, in the lee of the ledge on which Andrée's remains were first found, he discovered what appeared to be a stack of thawing reindeer hides. Slowly, one by one, he peeled them back. Underneath was the missing part of the puzzle, the remains of Knut Fraenkel, except for the backbone, thighbone, and pelvis.

By the time Stubbendorf returned to Sweden, Andrée's notebooks were being carefully transcribed and analyzed. They had survived the thirty-three winters remarkably well. Strindberg's notes were added to round out the story of their ordeal, complete except for the last few days.

On the afternoon of their takeoff from Danes Island, they made a brief entry and recorded the loss of their draglines. They made new ones that proved to be too light and too short. So, from the beginning, the *Eagle* was without much of its steering ability. Nevertheless, the men were in high spirits that first afternoon aloft. They drank ale, watched the panorama below, and spliced some ropes. Strindberg took some pictures before a light fog closed in during the late afternoon. As it became thicker, Andrée dropped ballast and the balloon rose into clear sky at twenty-two hundred feet. From this height they began to float down slowly, as tiny imperfections in the bag allowed the expanded hydrogen to seep out. They took turns sleeping, made sun observations to find their position, and dropped the second message buoy.

Early the next day Strindberg noted a cloud mass rising ahead of them. They dropped more ballast, but the balloon only rose to fifteen hundred feet, not quite high enough to clear the cloud bank. The aerostat settled slowly into the front and, as the sun's heat was lost, the gas cooled rapidly. Nothing was seen but the gray sea hidden from their view, here and there, by the motionless banks of heavy fog. They flew through miserable weather, half-mist, half-drizzle, that affected their spirits strangely. "We are gliding along so lightly and mysteriously," wrote Strindberg, "that we find it difficult not to hold our breaths for fear of upsetting the equilibrium of our feather-light balloon." In an hour the *Eagle* dropped to within a hundred feet of the frozen sea. More ballast was released before the balloon checked its descent.

At breakfast all three men were scanning the ice fields. The fog broke long enough for Strindberg to take another sun reading. All were elated when he announced they had traveled an incredible 250 miles northeast from Danes Island. "We'll reach the Pole yet!" Andrée declared confidently, but even as he spoke a fine rain began to fall. It soon froze on the great bag and began to weigh it down. The men emptied the last of their ballast bags and tossed their rope supply overboard. This checked their descent again, but that afternoon a grappling hook was thrown out when their craft started to settle once again. It continued a slow, downward drift. A few hours later, when their basket was only a scant few yards above the ice, Andrée discarded the polar message buoy (the largest, and the one intended to be dropped at their closest point to the Pole.) It did not help; the basket touched the ice and dragged along. Andrée dropped a heavy tarpaulin over the side and the balloon lifted twenty feet. The men cheered.

By evening, however, the situation fast became critical. Little else remained that they could afford to lose in the way of ballast. The fog continued to freeze on the dome and huge icicles hung from the balloon's perimeter. Again the gondola slapped the ice. It scraped, bounced, and bumped along the rough frozen surface. Slowly, the great gas bag shrank to a flabby shapeless thing and settled on the frozen pack. Strindberg estimated they had traveled 115 miles west of their last position.

By noon of the third day, the sun expanded the balloon's remaining gas and melted the ice from the bag. They were soon airborne again and, as they ate lunch, Andrée freed the homing pigeon that made its way to the *Alken*. About three o'clock in the afternoon, fog again shrouded them, ice re-formed, and the gondola resumed its bumping along the frozen ocean. In desperation, Andrée cast a medicine chest overboard. Abruptly, and to their complete amazement, the aerostat floated up to two hundred feet. Its canvas sails filled with wind and its makeshift draglines worked as they were designed. With a west wind, they now retraced their course of the day before, then a south wind pushed them northward. But if the intrepid Andrée still had hopes of reaching the Pole, he abandoned them that evening when the balloon sank again to within a few feet of the surface. The fog became thicker, the night colder. Sheet ice alternately formed, cracked, and re-formed on the dome. The men dropped two of their draglines without effect. When the third line was snagged in the ice early on the fourteenth and had to be cut free, they all new the balloon voyage was over. Reluctantly, Andrée selected a safe stretch of ice and opened the gas valve. The hydrogen hissed from the escape port and the *Eagle,* which had carried them five hundred miles in sixty-five hours and thirty-three minutes of flight, slowly settled onto the ice. They were 540 miles short of the Pole.

They planned their next move as they ate lunch on the ice. It was too late in the season for a rescue ship to reach them; next summer would be the earliest. Meanwhile, they would have to head for winter quarters at one of the depots Andrée had stocked for just such an emergency as this. The depots were on islands strung southward from Cape Flora to the Franz Joseph Group back to Danes Island. The expedition was three hundred miles from their departure point, but the nearest land depot was only 175 miles away at Ross Island. For some reason Andrée rejected both Ross and Danes Island and decided to strike out for Cape Flora, three hundred miles southeast. The men were grimly silent as they unloaded the gondola. Each reflected on his chance of reaching land; they all knew the trek ahead would be long and dangerous.

The party was fogbound for a week, during which time they loaded their sleds for the long pull. Finally, eleven days after they lifted from Danes Island, they started to pull their sleds across the broken, semifrozen ocean. They soon discovered that the surface was far more treacherous than they had estimated from the air. The summer sun had melted the ice unevenly and there was slush and large open lakes of water. The ice pack, which looked solid enough, crumbled under the weight of a man and sled. Leads of water, shallow and deep, had to be crossed. The morale of the men was high, however, and their health still excellent. With adequate clothes for warmth, a stout tent, and waterproof balloon cloth, each man pulled his 450 pounds cheerfully.

Before the first day had passed, they decided it was easier to travel the lakes and leads than to drag their sleds over the rough ice fields. Using ice floes as rafts, they ferried their sleds across the water. When their watches told them it was evening, they raised their tent, ate supper, and slept. As much as possible they lived by hunting bear, sea gull, and walrus to preserve their stored food. In the late evenings, which were still bright with the Arctic sun, Strindberg wrote notes to his sweetheart as Fraenkel wandered about to collect bits of frozen leaves and plants.

When Andrée set the course, he did not reckon for the unconquerable ocean current that lay between them and the nearest land. They struggled across the floating ice pack, pulled sleds that were overloaded, ferried them across dangerous leads of ice water, killed and ate game and trudged their way laborously over hammocks. They were blinded by fog and plagued by rain and soft, deep snow. These Andrée learned, were the summer conditions on the drifting ice floe.

After the fifth day of steady, backbreaking travel, Strindberg announced they had covered only two miles to Cape Flora. While they had been struggling foot by foot southward, pulling and ferrying their cumbersome sleds over the windswept hammocks, the huge drifting ice mass on which they were traveling had drifted them almost as fast northwestward.

Andrée called a council. The situation was serious and highly discouraging. With the Arctic winter night coming on, they were in danger of not reaching a winter quarters at all. Strindberg, who was guiding one sled, let it slip into a deep pool of frozen snow and slush. Bread, sugar, and biscuits were lost. This meant stricter rationing at a time when their meat supply began to dwindle. They decided to discard some more equipment to lighten the sleds for faster traveling. They cut down on their sleeping and eating time. This helped very little. The drift continued to cancel out their progress. During the first few days of August they actually lost ground on their giant treadmill.

Andrée was forced to make another decision. On the fourth, he announced they would abandon their efforts to reach Cape Flora and head instead to Ross Island. Abruptly, the drift stopped and reversed itself. Within the week they covered more ground than during the earlier two. On August eleventh they thought they had sighted land because they saw a higher elevation to their right. But it was only a great rise of ice on the floe.

When the sun dipped below the horizon for a few minutes on August thirty-first, they knew the Arctic winter was at hand. For a while there would be night and day; then, just night. It would get bitter cold, but now they were hopeful of reaching a safe haven in time.

By early September they had covered 150 miles, but the terrain and drift of the pack was causing their path to run due south, between Ross Island and Cape Flora. Again Andrée changed his plans, this time striking out for Northeast Land in the Spitsbergens, seventy-five miles southwest. On September third, the ice channels had grown larger as they drifted toward the Seven Islands. For a time they were able to use their canvas boat, but after only five hours they were forced to drag their sleds across the ice fields again.

Their extra efforts to get settled down before the Arctic night would overtake them was now hampered by sickness and an injury. Andrée wrote in his diary, "Fraenkel fell into the water today, and has diarrhea, and Strindberg has a pain in his foot, and I have diarrhea, but we covered a good distance today in any case." They discovered that seal blubber and the fat of bear meat eaten raw was effective in reducing their intestinal disorders.

Now a week-long howling wind upset their plans by pushing their ice island well away from Northeast Land. The fickle floes shifted unpredictably back and forth. They cracked and groaned; the leads became rushing torrents. Two sleds were damaged and the pain in Strindberg's foot worsened until he could barely walk. The men knew what had to be done; they would dig in where they were and pray their floe would reach land somewhere before the ice cap froze solid. They set to work building a foundation for a snow hut and on September seventeenth sighted a small land mass. It proved to be White Island. At first their floe moved toward it, but when it changed course their spirits sank. They drifted slowly past the island, helpless to change their direction. Fortunately, the currents came to their rescue and the next day they were drawn into the south shore of the rock-strewn haven. The island was little more than an immobile iceberg—its only advantage to protect them from the relentless polar drift of their floating ice pack. By now, however, their snow hut was finished and Andrée turned more to his record. He wrote on September twentieth, somewhat ominously, that he was annoyed because the spare parts for his Primus had been accidently left behind in Spitsbergen. Apparently, it was not working well. But they felt so secure, so warm and comfortable, so amply stocked with game and birds, that when their chance came to go ashore for the winter, they elected to remain on the pack.

Early in the morning of October second, fate changed their minds. The low rumblings of the ice pack grew into an ear-shattering roar. The floe trembled and broke into many smaller floes as a huge water-filled ravine opened beneath their hut. One wall disappeared into the swirling torrent as water rushed in to soak them. Late in the day of October fourth, they decided to move ashore. This was done during good weather of the fifth and sixth.

Andrée's last diary entry was dated October eighth. It was optimistic and routine, with no suggestion of danger or threat of illness. There was not even a premonition of trouble. The cause of the tragedy would have to be found at the campsite, by the skeletons of the dead, and the remains of their handiwork. Toward the end of the diary, Andrée's handwriting is smudged and illegible. Although thirty-three years of exposure to the elements damaged these final pages of narrative, the explorer had left in them little in the way of information. What happened after the trio reached the shore isn't clear; from the clues at their final campsite we can only surmise what followed.

Even if the notes had been complete, they still would not have revealed how Andrée and Fraenkel died, though possibly how Strindberg did. We know he was buried by the other two, that his foot was injured, and that he probably perished within a week and a half of the party's arrival on the island. Did he catch pneumonia? Did his injury and recurring diarrhea so weaken him that he died? How long did Andrée and Fraenkel live after they buried their comrade under the stone cairn? A day or two at the most. Conditions showed their campsite to be only partly set up, with two sleds yet to be unpacked. At this late date in the season, Andrée and Fraenkel must have been busy getting settled down for the winter. If Strindberg died on October ninth, it was likely that all three sleds were still largely loaded, because of the limited hand-over-hand relay that brought their gear to their campsite from the beach. Death interrupted them.

The one unloaded sled that was found gave a clue to the likely cause of Strindberg's death and suggested that he died accidentally soon after Andrée's final diary entry on October eighth. Fraenkel's sled was still loaded, as was Strindberg's, which was alone, farther from the tent. But Andrée's sled, near the tent site, had been completely unloaded—and it had been done hurriedly. A tarpaulin, tied across the top to protect its contents from the drifting snow, had been torn off and the remaining contents were dumped, helter-skelter, on the ground. But why?

Strindberg probably died by drowning, within shouting distance of their camp. He had ventured back on the ice pack for some reason, perhaps to recover an object they had dropped. It was a dangerous time of year to walk on the floe; fresh snow hid the weak patches, and when Strindberg suddenly fell through the ice, his comrades likely saw him floundering or heard his cries. They were unable to approach close enough to reach him, because the same weak ice that had broken under Strindberg now threatened to collapse under them. They needed a rope, or something that would distribute their weight on the treacherous ice. A sled! Andrée and Fraenkel rushed ashore and overturned the first sled they reached—Andrée's—scattering its load in a pile. This was how it was found when the camp was discovered.

But they returned too late; Strindberg drowned quickly. They used the sled to bring his body to the camp and probably buried him the next day. The summer of 1897 must have been warm on White Island. Only during warm summers would the extreme shore edges have exposed the rocks and stones they gathered to heap over Strindberg's remains.

But what of Andrée and Fraenkel? They did not starve to death. They were on an island abundant with game and fish; they had rifles and ammunition. The remains of a bear they shot was a few yards away. They shot and ate birds—guillemots—and their sleds still held a good store of canned and preserved food. The parts of their bodies that were recovered bore no traces of injury. They had not been attacked. They did not die of exposure. Certainly they did not freeze to death. They had more than enough clothing, plus an oversize sleeping bag. True, they were lightly clad when found, considering the weather, but more clothing was lying unused within arm's reach as they lay in the tent. Why didn't they use the two bearskins as rugs to lie upon in the tent? Why were they *outside* their sleeping bag? Why did the third sled remain packed, with some of the expedition's most important gear—as though in preparation for a journey?

It's quite possible the deaths of Andrée and Fraenkel came during the evening of the day they buried Strindberg. All investigations indicated they had not occupied their camp long. Fraenkel wore a woolen undershirt, a flannel outer shirt, a vest of brown cloth, a chamois leather vest, a woolen jersey, and a cloth coat with a fur collar. Fraenkel had put on a black mourning ribbon. He wore only socks; Andrée wore shoes.

There was doubt as to how the lower part of Fraenkel's body was dressed, for it seems a bear tugged at his body and pulled away his legs. This must have happened the year after his death, for a beast powerful enough to tear away a man's legs would have dragged away the entire body unless it were frozen solidly in the ice. But ice would not have formed until after a summer thaw followed by a freeze.

Fraenkel's remains were in the middle of the tent. Behind them, close to the northeast wall, lay their sleeping bag of reindeer skin, wrinkled and frozen. Andrée's remains were on the low ledge of rock, just above the sleeping bag. The two had therefore died within the tent, on its fabric floor, separated by a scant few feet. If either had died sometime before the other, he would have been buried by the survivor. Apparently, the tent was warm and closed and there was no need to dress more warmly or get into the sleeping bag. Andrée and Fraenkel sank into their last slumber, side by side within the tent and, from all indications, within minutes of one another.

In the final analysis, we find there was no single factor involved in their deaths—but, rather, two.

In 1952, a Danish doctor, E. A. Tryde, published a book on

the Andrée expedition after he had carefully studied the notes and references Andrée had made on their trek across the ice. Early in the journey, Tryde discovered, they had suffered from fever, cramps, diarrhea, swollen and painful eyes, stomach pains, muscular pains, rashes, and small boils. These symptoms indicated their illness was trichinosis, a parasitic disease usually associated with pork, but here caused by eating infected bear meat. The disease has been found to be sporadic among certain marine animals and endemic in Eskimo dogs and polar bears. Andrée, Fraenkel, and Strindberg had probably been suffering from the tiny trichinae, which can infect man and even cause his death.

In the Andrée Museum at Grana, Sweden, Tryde discovered the preserved remnants of two polar bears shot during their journey. Microscopic examination revealed that both animals had been infected with the trichinae. According to Andrée's notes, they had eaten the flesh—often raw—of more than a dozen bears during their stay on the ice. There was good reason to believe that at least some of the bears they had shot earlier on their trek were also infected. Millions of trichinae must have invaded the bodies of the three men shortly after they landed in the *Eagle*.

Trichinae seek out the transverse muscles. Although the heart has musculature of this kind, the tiny parasites do not like its pulsating contractions so they move on through the chest. But their passage through the heart leaves behind a network of tiny pinpoint scars. Tens of thousands of trichinae must have passed through the heart muscles of Andrée and his companions. A victim suffering from severe trichinosis usually dies from general exhaustion, or pneumonia, or from sudden heart failure brought on by great exertion. Strindberg's thrashing to save himself may have caused his weakened heart to fail; and later, as an exhausted Andrée and Fraenkel tried to rescue and then buried Strindberg, the disease worked—in concert with another factor—to kill them. But what was the other agent in their death?

Like all inexperienced white men, the Andrée party made their campsite in the lee of the cliff as a shelter to the winds—from at least one direction. Eskimos would never consider placing a winter campsite in the lee, because the same obstruction that shelters the camp from the wind also shelters the drifting snow and a snowdrift will form there. If the lee is high enough—the cliff towered thirty feet—and the storm lasts long enough, the snow would completely bury the camp. The snow blanket could cause death from suffocation or lack of ventilation or from the caving in of the shelter under its weight.

It appears that Andrée's favorite toy—his Primus stove—was primarily to blame for their deaths. Sometime after the two had laid Strindberg to rest, they prepared to settle down for the night. Both were weary, suffering from the effects of a disease of which neither was aware. The temperature outside was not very cold as the winter's first snow fell. The blanket of white that covered the tent also insulated it. Their tent was large, made of varnished balloon cloth and, as the interior became warmer, while the evening meal cooked on the Primus stove, Andrée and Fraenkel removed their heavy coats. After eating, they rested and talked.

When fuel combustion is incomplete, deadly carbon monoxide gas forms. It has no odor and gives no warning. Their sealed polar tent did not allow the heat—and fumes— from the Primus to escape. Death by carbon monoxide poisoning is a hazard in the barren Northlands. William Barents, who discovered Spitsbergen in 1594, was wintering on Novaya Zemlya in 1596, when several of his men fell unconscious in the hut. They were burning coal for fuel, and in the unventilated room carbon monoxide overcame several of his companions, who narrowly escaped death. In April of 1911, four members of the second Stefansson expedition, who were camped on the ice of Coronation Gulf, also narrowly escaped death from the carbon monoxide fumes of their Primus stove. Only at the last second did Stefansson realize what was happening in time to extinguish the flame. In August of 1938, when Admiral Richard Byrd was at Little America in the Antarctic, two of his men found a companion unconscious in a back room heated with ordinary kerosene. The flue pipe had become clogged with snow from a blizzard. In all three cases, imperfect combustion combined with no ventilation allowed the accumulation of deadly carbon monoxide gas.

Andrée and Fraenkel died warm, comfortable, and without foreboding, slipping gently into the slumber that preceded death. That both were undoubtedly asleep was evident from the condition of the stove.

When the camp was found, the Primus stove was standing between the bodies, and still half-filled with kerosene. Dr. Gunnar Horn distinctly remembered that, when he found it, the air valve was closed. This meant the men had died when the Primus stove was burning. With the air valve closed, the stove would continue to burn until the air pressure inside, built up by the small hand pump, was depleted. Then it would extinguish itself and the cold of the Arctic night would come.

Andrée was well aware of the hazards of remaining in a closely confined tent with poor ventilation. It is difficult to imagine that alert men in full possession of all their faculties would not act to save themselves as soon as they had the first indication of asphyxia.

But who can say for certain that this is the way it happened? Of the many theories, it seems the soundest.

Sweden paid a final tribute to the explorers. On September 30, 1930, seventy-five thousand people filled the harbor city of Göteborg. Thirty-three years earlier the *Svenskund* had escorted the three explorers to Danes Island. Now refitted, it carried the whitened bones of the aeronauts home. All along the coast of Norway church bells tolled at their passing. The floating cortège continued to Stockholm escorted by Navy vessels, artillery salutes, and the fluttering fall of wreaths from airplanes overhead. Down the streets of Stockholm in solemn mourning, the remains of Andrée,

Fraenkel, and Strindberg were carried to the Church of St. Nicolas. Thousands of their countrymen, in company with Gustavus V, attended the service. When the coffins were taken to their burial place, the king spoke these words:

> In the name of the Swedish nation I greet the dust of the Polar explorers who, more than three decades since, left their native lands to find answers to questions of unparalleled difficulty. A country's hope to be able to honor them in their lifetime after a successful journey was disappointed. We must submit to its tragic result. All that is left us is to express our warm thanks to them for their self-sacrifices in the service of science. Peace to their memory!

7
Somewhere at Sea

The early over-water flights carried neither wireless nor flotation gear, so the clues most needed to explain the loss of planes at sea—position reports and survivors—were frequently missing. In most cases the question of what happened out there will never be answered.

Early radios were heavy, cumbersome, and undependable. Transmitters lacked range; receivers were subject to whims of the weather. Many an Atlantic air adventurer believed his chances as good without them, especially when he plotted his course off the established sea lanes, anyway.

Anyone can theorize on the causes of air losses over the North Atlantic between 1919 and 1931. Speculation is cheap, especially with the cockpit full of ghosts that haunted the early daredevils. The weather was a gloomy, ever-present threat. Forecasts were little better than guesswork. Ice could choke the engine and destroy the airflow over the wings and propeller. Snow could weigh down machines already overloaded with fuel, and there were no devices to eliminate these hazards. No plane could fly above the weather as they do today; pilots had to outwit the elements, endure them, or fall victim to them. Even in clear weather, fickle winds from unexpected quarters at unexpected velocities magnified the problems of steering with an erratic magnetic compass. Looking back, it seems miraculous that a few determined men and women survived the hazards and ran the gauntlet unscathed.

Here are the stories of some of those who did not.

The first attempt to fly the Atlantic in a heavier-than-air machine was made by Harry Hawker and MacKenzie Grieve. On May 18, 1919, they took off from Newfoundland for the fifty-thousand-dollar prize offered by the *London Daily Mail*. Their Sopwith biplane, powered by a 375-horsepower engine, was forced down in mid-Atlantic, and when it failed to arrive in England, the worst was assumed. Then, seven days later, the Danish cargo boat *Mary* docked in England and lashed aboard it was the missing plane. The two fliers were safe, fished from the water as they clung to their floating aircraft.

The White Bird

May 8, 1927. Joan of Arc Day, Le Bourget Field, outside Paris. In the pale dawn, mechanics pushed a ghostly white biplane from a darkened hangar. Thousands of spectators, many of whom had been waiting for this moment since midnight, broke into cheers. Today the rash Nungesser and his one-eyed navigator, Francois Coli, were going to fly to America.

The pair stood to gain fame in one of two ways if they were successful. A new transatlantic record, or if they landed at New York, the twenty-five-thousand-dollar Raymond Orteig Prize. News of their stepped-up departure date traveled fast. When Nungesser learned that a little-known airmail pilot by the name of Charles Lindbergh was ready for a New York takeoff at any moment, he moved his departure date ahead two weeks.

To the average Frenchman, Nungesser was surrounded with an aura of supernatural invincibility. They called him "Nungesser the Indestructible." And well they might. He had emerged victorious from forty-five air-to-air combats over the western front; he survived seventeen crashes that broke almost every bone in his body. His entire lower left jaw, palate, and shattered left leg were reinforced with platinum plates. Time and again his poorly mended bones had to be reset. Certainly he was the most daring of the French aviators and the most determined. When Réné Fonck, with whom he had competed for France's top war victories, failed in his attempt to fly from New York to Paris, Nungesser convinced himself that he would succeed.

After the war, Nungesser operated a flying school, then barnstormed the United States. Late in 1926, he announced his plans to cross the Atlantic by air, not west to east to take advantage of the prevailing winds, but east to west—the hard way. As he boarded the boat that was to carry him back to France, he bid his brother good-bye. "Farewell, Robert. I'm going home the slow way, but I'll be coming back the *fast* way—by aeroplane!"

The air hero set to work. He wanted a large monoplane with an enclosed cabin. Fonck's plane was like this, but when news came that he had crashed on takeoff and two crew members had perished, Nungesser changed his mind. Next he considered a small, light airplane to carry one man and much fuel, a flying gas tank. The government refused to subsidize him so he went to the manufacturers. Their assortment of planes was anything but encouraging. Finally, he decided upon the great Levasseur P.L.8 biplane, then under manufacture for the French Navy as a carrier-borne three-seater patrol bomber. Nungesser had it modified to suit his needs. A 450-horsepower Lorraine-Dietrich twelve-cylinder engine was installed and extra gas tanks were fitted. In fact, the entire forward part of the fuselage was one great fuel cell. The fuselage was reworked into two open cockpits (Nungesser decided to carry a navigator) and the landing gear was designed to be dropped after takeoff to lighten the machine, give it less drag, and prevent it from flipping over when it landed in New York Harbor. The propeller could also be locked in the horizontal position for the water landing. The bottom of the boatlike fuselage was shaped with a planing bottom and contained watertight compartments to keep the plane afloat in case of a forced landing at sea. To aid in immediate recognition should such an emergency occur, the entire plane was painted a chalky white.

Each side of the fuselage bore the macabre insignia Nungesser painted on all his planes—a large black heart inside of which was a skull and crossbones. Over the skull was a coffin. He considered this his good luck charm as well as his trademark.

When Nungesser and Coli appeared, riding toward the waiting plane in an open car, the cheers turned to a thunderous roar. The car stopped. A young girl stood on tiptoe and threw Nungesser a rose. He caught it in mid-air, touched it to his lips, and blew the girl a kiss. Then, erect, but with a slow limp, he walked to the plane, Coli trailing. The mechanics topped the tanks as the two fliers watched, and 880 gallons of gasoline were ready to be lifted into the air by the clumsy-looking biplane.

Captain Coli, accustomed to the precise workings of ocean navigation, was less devil-may-care about the venture. Nungesser had not taken time to test the Levasseur with its full load of fuel, so neither knew for certain if it would get off in the half-mile of runway. The ace refused to worry about it. "We'll either make it or we won't," he said lightly, and dismissed the question.

Coli was still uneasy. His calculations could only be based on the *assumed* wind conditions over the Atlantic, and they expected to buck head winds all the way. The normal cruising speed of the patrol bomber was about 120 miles per hour. It could stay aloft for forty-two hours, but its ground speed might vary from 80 to 120 miles per hour, depending on the wind direction and velocity. At the very best, Coli could afford only a two-hundred-mile leeway. This was cutting it close, but the nonchalant Nungesser accepted it cheerfully. The navigator shook his head. Everything, he knew, would depend on Nungesser's flying ability and his own navigating skill. And the weather . . .

The plane carried no radio; it had been removed a few days earlier. There were no life jackets or rubber dinghies, no survival equipment. Coli, long experienced in the fickle ways of the sea, protested, but Nungesser replied coolly, "The idea, Captain, is not to swim to America, but to fly there. If we have to land on the water, the fuselage will keep us afloat until we're picked up.''

But Coli knew the sea, especially the North Atlantic. If the waters were rough, *l'Oiseau Blanc* would sink with its crew in a matter of minutes.

By any standard it seemed almost unbelievable that an experienced pilot and navigator would attempt to span the Atlantic under these uncertain conditions. The only encouraging word was the weather office's last-minute report that, miracle of miracles, they could expect a tail wind for the first third of their journey. Nungesser was elated; Coli felt better.

The two men, dressed in yellow flying coveralls, climbed into the cockpits, where their carefully weighed provisions —canned fish, bananas, sugar, chocolate, coffee, and brandy—had been stored. Nungesser looked over the crowd. There was Maurice Chevalier, Georges Carpentier, and the incomparable Mistinguett, France's darling. On the fringe of the crowd stood General Girod, Nungesser's wartime commander. Nungesser was pleased; all France was there to see him off; the workers, the *artistes,* the military.

The engine roared to life. At five-seventeen, *The White Bird* lumbered reluctantly forward. It seemed impossible that the five tons of airplane and gasoline could become airborne in Le Bourget's twenty-five hundred feet of runway. The tail lifted sluggishly, then slowly settled back down. A cry of dismay swept the crowd. *"Mon Dieu!"* cried Girod. "He'll never get off the ground!"

The tail lifted again as the white, straining machine rolled faster. Nungesser swept past the halfway mark of the runway; his main wheels still had not broken ground. Then they lifted slightly and settled back. Less than five hundred feet remained and the overloaded plane made every effort to fly. In the cockpit, Nungesser saw the end of the runway looming up fast. It was too late to stop now and he knew another backward pull on the control stick to try to force the lumbering plane to fly would be disastrous. His bird would stall and crash as had Fonck's a few weeks earlier.

As the final yard of runway slipped under the wheels, the huge biplane staggered into the air and held. Its engine labored full out as the wide wings reached out for enough air to begin the climb. The crowd went wild. Nungesser was holding fast; his uncanny luck was still with him.

Lifting itself by inches, the plane headed toward the sea. It was escorted by other planes as far as Cape Etretat, north of LeHavre, which it passed at six forty-eight A.M. Nungesser and Coli were last seen, officially, five hours later as they left the Irish coast, heading into the teeth of a strong Atlantic wind.

The next day a false report of their successful landing in New York Harbor stirred France to a brief celebration. But when the truth became known, the celebration turned to a riot. In Paris, indignant Frenchmen burned stacks of newspapers that carried the unconfirmed report. Editorial offices were in danger of being sacked and pillaged. Windows were smashed.

Hours passed and no word came from any quarter, ship or land, of a sighting. It was obvious that the daring pair had come down somewhere in the North Atlantic. On May tenth, they were officially reported missing.

Despite the many theories as to where Nungesser and Coli went down, it was never disproved that they might well have crashed only minutes from the shores of France. There was, however, a fairly accurate report that a plane resembling *The White Bird* was sighted off the Irish coast, well out to sea at five after ten on the morning of May eighth. Considering the plane's cruising speed and the elapsed time from Le Bourget, officials decided "this could have been *l'Oiseau Blanc.*"

Beginning at midday of May ninth, conflicting reports appeared. News agencies claimed Nungesser and Coli were indeed sighted over Cape Race, Newfoundland, and over Newbury, and Boston. New York papers were unable to get confirmation. Had a plane actually passed over these points? Yes, one had, but it was not *The White Bird*. From New York to Halifax, observers scanned the skies, and despite mist and fog, a squadron of U.S. military planes set out from Boston to meet the Frenchmen. They returned without making rendezvous.

On May twenty-fourth, a trawler that had been in the Atlantic at the time of the flight, reported seeing a smaller vessel towing a white hydroplane. Nungesser and Coli? The British Admiralty promptly cleared up the matter; one of their planes had been forced down at sea. In the wake of this, two other ships, the *Bellaline* and the *Dana*, each reported having sighted the missing plane not far from the American shore. An investigation showed they had indeed been on the route plotted by the fliers, but at the time the sightings allegedly occurred, the biplane had long since exhausted its fuel.

Did *The White Bird* come down on land or sea? What about Labrador? Did the pair crash somewhere on its frozen wastes? Then came a report from that very place. Trappers saw rockets in the sky one evening about the time of the flight. The lights had been some distance from their camp and, on searching the area the following morning, they found nothing. From Newfoundland came other reports of an airplane engine heard overhead when it was certain no other plane was flying in that region.

Canada? What of its sprawling forests and snow plains? A Canadian trapper claimed he found a written SOS message signed "Nungesser." In fact, he produced it. It said Nungesser and Coli had come down in the far north. In Paris, Nungesser's mother examined the handwriting and thought it like her son's. But because of the poorly worded text of the message she said, "It seems unlikely that Charles could have written it."

There are Frenchmen who believe today that Nungesser and Coli still live. Like their faith in the invincibility of Guynemer, to them these fliers will never die. Frenchmen may admit their heroes crashed, but only after they reached America. They came down in the Canadian north where they live on today as captives of a remote Indian tribe. Others believe they died of starvation; or that faulty American weather reports caused them to crash within sight of land. These beliefs were nurtured by messages found later in floating bottles. One turned up on the Dutch coast in 1929. Another was found in 1933, and the last one was picked up in 1934. All were cruel hoaxes played on a sympathetic public by mentally deranged persons.

Perhaps Nungesser and Coli almost made it to America. Early in 1961, off the southwest coast of Maine, a Casco Bay lobster fisherman pulled up one of his pots. Caught on the side of it was a jagged piece of metal wreckage. It appeared to be a riveted scrap of aircraft aluminum. The man scraped away the slime accumulated during its many years on the ocean bottom and found it had once been painted white. A remnant of *l'Oiseau Blanc?* Who can say?

Two weeks after Nungesser and Coli disappeared, Charles Lindbergh made the first nonstop solo flight from New York to Paris. Somewhere along the way he may have passed near the place where the brave Frenchmen faltered and fell. As other successful transatlantic flights followed, the Nungesser-Coli tragedy gradually faded from the public mind.

How final were the words of Captain Venson, one of the pilots who escorted the intrepid airmen to the cliffs of Etretat, when he described his last view of *The White Bird*.

"She gradually vanished ahead of us in the opaque milky haze tinged with red . . . as day rose."

The *Saint Raphael*

A feverish, almost compulsive spirit caught hold of airmen when they contemplated the Atlantic. The temptation refused to be stilled. Nungesser and Coli were missing only three months when the beckoning finger of Dame Fortune lured another plane over the sea, and the Atlantic took its toll again. This time it was two men and a woman.

On August thirty-first, British fliers Colonel Frederick

Minchin and copilot Captain Leslie Hamilton took up the challenge to make the first east-west crossing. In an effort to succeed where the Frenchmen had failed, they decided to reduce the flying distance by starting from Cornwall, England. Their passenger was Princess Loewenstein-Wertheim. She was not a young woman, one who could take in stride the rigors of thirty-six hours over the Atlantic in a cramped cabin. She was sixty-two, but she wanted to be the first woman to cross the ocean by air. As the principal backer of the flight, she had no trouble making the arrangements, and the plane was equipped with a wicker armchair and an army cot to provide her the most comfort under the confining conditions. Outfitted in a blue leather flying outfit and suede boots, she traveled light, taking two handbags, a basket of food, and two hatboxes.

At Upavon, a small flying field on Salisbury Plain, the mechanics finished their last-minute check of the nine-cylinder, 450-horsepower engine that was to power the blue and yellow Fokker monoplane thirty-six hundred miles to Ottawa. They stood back and waited, but the trio gave no sign they were ready to go.

A light, steady drizzle was falling when a car drove onto the muddy field, carrying the Roman Catholic archbishop of Cardiff and his acolytes who, at the princess's request, were to be in attendance. The archbishop blessed the plane, the *Saint Raphael,* and sprinkled it with holy water as the princess bowed her head and prayed silently. The ceremony was short, farewells were quickly said, and within ten minutes the Fokker lifted itself without difficulty and disappeared low into the ragged wispy clouds.

Over the water, Minchin and Hamilton settled G-EBTQ onto a steady course for Newfoundland. Droning along at one hundred miles an hour, the six-and-a-half-ton aircraft handled well with its eight hundred gallons of fuel. Although the weather was squallish, the monoplane alternately broke into clear patches of air over the churning whitecaps, then into low-hanging "scud" again. They carried no wireless equipment, so their only contact with the world below was a visual one.

Occasionally, they sighted a fishing boat or a freighter near the coast, but after an hour the surface vessels thinned out. Several hours later the *Saint Raphael* was sighted well out to sea by a lone tanker that radioed the Fokker's position to London. It proved to be the last contact with the plane, for it was never reported again. After forty-five hours, hopes faded in Ottawa.

Then a curious phenomenon was reported on the night of September third, when the Fokker's fuel tanks would have been empty over twenty-four hours, something that makes all the more puzzling the record of the lighthouse keeper at Belle Isle, Labrador. So rarely is the monotonous splash of the sea broken by an alien sound that this man could not have imagined what he saw and heard. Shortly after midnight, he saw a red light in the southeastern sky, with a steady white light behind it, obviously the navigation lights of an airplane. Since aircraft carry a red light on the left wing and a white light on the tail, the man was looking at the left rear of the plane.

"Then I heard the engine," he said, "and watched the lights until they disappeared in the vicinity of Battle Harbor about twelve-thirty A.M."

Every attempt to track down the identity of the midnight flier ended in failure. Just who—or what—was moving over the barren stretch of isolated loneliness was never discovered.

A few days later, scraps of a broken airplane were found off the coast of Iceland. Among them was a wicker chair. There was no doubt this was all that remained of the *Saint Raphael,* despite the fact that the bodies of the crew and passenger were never recovered. Thus, instead of being the first of her sex to cross the Atlantic by air, Princess Lowenstein-Wertheim earned instead the dubious distinction of being the first woman to perish on a transatlantic flight.

Old Glory

Lloyd Berthaud, one of Columbia Aircraft Corporation's long-time pilots, planned the next transatlantic flight. He proposed to fly nonstop from New York to Rome, a distance of 3,850 miles. His plane, *Old Glory,* was also a single-engine Fokker monoplane with a 450-horsepower Jupiter engine.

On September 7, 1927, the long-awaited news flash came. Berthaud had hopped off from Orchard Beach, Maine, and was on his way. With him was his navigator, J. D. Hill, and Philip Payne. Payne was managing editor of the New York *Daily News,* one of the directors of the Hearst chain, and was personally selected by William Randolph Hearst to accompany them.

The *Old Glory* was better equipped than most of the early transatlantic contenders. It carried a wireless as well as a newly developed automatic apparatus to send continuous distress signals in case of a forced descent at sea. The radio transmitter worked so well that soon after takeoff the steamer *Berlin,* twelve hundred miles out of New York, picked up clear signals. Another innovation was a device for distilling seawater. Well-selected provisions and an inflatable rubber dinghy were stored in the fuselage, and in the rear of the cabin was a wreath. On its ribbon was written: TO NUNGESSER AND COLI—YOU SHOWED US THE WAY; WE ARE FOLLOWING. Although the gallant Frenchmen flew in the opposite direction, the sentiment was understood. "We intend to drop it in mid-Atlantic in honor of the two French airmen," Berthaud told reporters before takeoff.

Their Rome destination allowed Berthaud to plot their course farther south than Lindbergh's. This enabled him and his crew to fly along well-traveled shipping lanes, which increased their chances of survival in the event of a crash at sea. Several hours out of his jumping-off point, Berthaud was seen by a number of ships that reported all was well. The Fokker plodded steadily on course. It passed within three hundred feet of the SS *California,* plowing along 350 miles east of St. John's, Newfoundland, and the plane's identification letters, WHRP, were clearly seen. Berthaud, it appeared, had a winning combination.

Then, toward nightfall, the seas turned heavy. By early morning, strong westerlies churned the waters into heavy swells. Wireless operators aboard ships whose course through the rough weather was close to *Old Glory* heard nothing but static. Bridge lookouts scanned the dark and blustery skies for the telltale colored navigation lights. But in the low cloud cover, nothing showed itself.

On board the *Transylvania,* the wireless operator kept close vigil. It was foul weather indeed, and he thought Berthaud and his crew would be having a rough time of it if they were within 250 miles of the ship.

He twisted the dials, picked up a time signal, scraps of a message from London, then turned back to the *Old Glory*'s transmitting frequency. Suddenly, out of the racket and noise, he heard the wavering dot-dash of "*Old Glory* calling. . . . *Old Glory* calling. . . ." Then came their position, followed, to the operator's dismay, by the emergency signal of the air: SOS. The garbled message was so broken with static he could only make out "Newfoundland east . . ." and their position of latitude 48:03 north and longitude 41:15 west. He scratched the message on his pad and raced to the bridge. The weather was growing worse.

When Captain Bone read the message, he ordered the ship to alter course immediately. Other vessels were already hurrying to the position Hill transmitted, for the SS *California* and the SS *Carmania* picked up the distress call simultaneously. They were converging at full steam on the *Old Glory.*

The *Transylvania* was nearest the transmitted position of the plane, and after six hours in the rough seas it reached the spot. But in the growing light of dawn there was no trace of the plane. Captain Bone quartered, zigzagged, and circled the area in a systematic pattern. Other vessels appeared on the dim horizon and joined the search. There was still no sign of the *Old Glory*. Another six hours passed without result.

There was no misunderstanding about the distress position Hill transmitted. At least not among the ships that heard it. Six in all confirmed it, converged upon it, and marked the area well, but once the section was thoroughly combed, there was no hope that the plane was still afloat or the fliers had escaped a watery death in their inflatable boat—if they had had time to launch it.

At twilight, after more than twelve hours of searching, the captains reluctantly turned back to their courses and continued to port.

Did the weather force *Old Glory* down? Or was it engine trouble, a broken fuel or oil line? The world wondered. Berthaud definitely was not lost; he could not have given a position report. But was he certain of his position when he transmitted his latitude and longitude?

In the weeks that followed, tankers, liners, and naval ships found scattered bits of floating debris that could only have been the remains of the Fokker. The wreckage was generally found one hundred miles north of Hill's position report. Thirty-four feet of wing was recovered, on which was painted the Stars and Stripes. Fuselage and tail fragments were also fished from the water. They were the first sizable parts of any of the overwater planes to be found, and they told the story only too well. Berthaud had been unable to ditch the monoplane successfully, and on impact with the mountainous waves it had broken up within seconds, splintering into hundreds of pieces. No one held hope of finding the crew, and no one knows why they were forced to ditch.

The wreath intended for Nungesser and Coli now marked two graves. All had perished in the same fashion—trying to span the Atlantic.

The *Sir John Carling*

The tragic end of Berthaud and his crew failed to dissuade two other Britishers, pilot J. D. Metcalf and navigator

Terence Tully, from making the transatlantic flight. The newspapers were still carrying headlines of *Old Glory*'s disappearance when they departed on September 1, 1927, from London, Ontario. Their destination was London, England. No one had yet flown there from Canada nonstop.

Their plane was the *Sir John Carling,* a Stinson Detroiter monoplane that was to carry them over Harbor Grace, Newfoundland, and thence to England. Bad weather forced the plane down at Caribou, Maine, and after other delays the latest Atlantic contender began its ocean crossing from Harbor Grace on September seventh. Then came the message that they had been sighted leaving the eastern shore of Newfoundland, on course. But when Metcalfe and Tully left the last bit of coastline behind and headed over the waiting Atlantic they disappeared forever. At Croydon Aerodrome outside London, the fog rolled in and the watch waited in vain.

Two days later, at Newquay, Cornwall, bits of the silver-gray wreckage washed ashore and were identified as parts of the wing and rudder of the *Sir John Carling.* The plane was expected to cross the Galway coast that Thursday afternoon about the time a strong gale hit the Irish coast. They had come down at sea, just short of the shore of Britain. The sea had won again.

A Gambler's Chance

A month later, a lieutenant-commander in the Royal Navy, who had served with distinction during World War I, made the next attempt. H. C. MacDonald was convinced he could fly from Newfoundland to Croydon in a tiny eighty-five-horsepower De Havilland biplane that was modified for an incredible thirty-six hundred-mile range. His announcement stunned veteran airmen. "Foolhardy. Absolutely foolhardy," was their reaction to the press. MacDonald, it appeared, had a total of eighty hours in the air, only a half-hour of which was night-flying experience. Some months earlier, with only sixteen hours to his credit, he flew to India. Remarkably, he arrived safely, but on his return flight from Baghdad he was forced down in the Sahara and was seized by hostile Arabs. Only the timely arrival of an Italian armored car saved him from death. MacDonald was thrilled; to him all life was a grand adventure, and the Atlantic only whetted his appetite for new experiences. Before he sailed for Newfoundland, his friends tried to dissuade him. All he said was, "I know I'm taking a gambler's chance."

At Harbor Grace, on the crisp morning of October eighteenth, MacDonald double-checked the instruments in his open-cockpit biplane. The small group of spectators had watched transatlantic hopefuls come and go, but never in a plane as small as this one. The tiny De Havilland's wingspan was only twenty-six feet. A few men shook their heads as they conversed quietly in small groups. MacDonald was ready; he stored some sandwiches and a thermos of coffee in the cockpit. Grinning, he said, "I'll finish these in London tomorrow." He climbed into the cockpit and signaled a mechanic to swing the propeller. The engine sputtered, coughed to life as its four exhaust pipes popped away.

The biplane was so burdened with its extra fuel tanks that the undercarriage sagged and the tires bulged. The mechanic helped get the plane moving, and MacDonald slowly taxied to the takeoff point. He turned his plane into the teeth of a twenty-five-mile-an-hour wind and paused briefly. Then the Gipsy engine roared full out, as full out as eighty-five horsepower can roar. With its exhaust crackling in the chill air, its pistons strained to get the airplane rolling. With a normal load, the light aircraft would pop into the air within a hundred feet, but now it took a lumbering run of six hundred feet before it staggered into the air.

MacDonald climbed slowly and steadily. He circled the aerodrome twice as he gained altitude; then, with a final dip of a wing, he headed over the water at eleven fifty-one. After that, he was sighted once or possibly twice as he flew eastward. Seven hundred miles out of St. John's, he was sighted by the Dutch steamer *Hardenberg*. He was airborne seven and a half hours. He was making good his planned track, was right on schedule, and was averaging almost one hundred miles an hour ground speed. MacDonald said before takeoff he would probably reach the coast of Ireland the following morning about seven o'clock. But without wireless, it was impossible for ship and shore stations to get a bearing on him to check his progress during the long night.

The British liner *SS Montclare* was steaming 150 miles off the coast of northern Ireland late in the evening of the second day, when several passengers saw a light moving in the distant sky. It was too far off for them to hear the sound of an engine.

In Kensington, London, Mrs. MacDonald, a slim brunette, waited anxiously with their five-year-old son Ian. The boy knew of his father's flight and happily told callers, "My daddy's in an aeroplane . . . over the . . . the Al . . . Al . . . anic!" Mrs. MacDonald put him to bed at seven, and continued her vigil.

"My husband had wanted to do this for a long time," she told a newsman. "He worked hard to make the arrangements, even had his machine dismantled and shipped to America in boxes." As she spoke, police from Malin Head to Dingle Bay on the west coast, and Cape Clear to Land's End on the southwest coast, were keeping beacon fires alight and ears alert for the De Havilland. The weather over England turned wet and stormy as the expected hour of MacDonald's arrival drew near. When eleven P.M Friday came and passed, hoped waned that the handsome naval officer would reach his goal.

Mrs. MacDonald refused to believe the worst. She clung fast to the hope that her husband had been picked up by a vessel at sea, for he had reminded her that Hawker and Grieve were missing seven days before they reached port. "Don't give up hope," he had said in parting. Then, two close family friends, Sir Herbert Barker and another, told her of their identical dreams. They each said they saw a rocky island about two hundred miles off the west coast of Scotland. Mrs. MacDonald hastened to check the charts. She said she believed the island was Rockall, and immediately contacted the Admiralty. The Admiralty was sympathetic, but carefully explained that Rockall was a small, high rock, cold and exposed to the sea winds. Anyone there could not survive long, they said.

Mrs. MacDonald was not convinced. She tried again by telephoning Croydon Aerodrome and asking for a plane to be sent there to search. But the Admiralty intervened and refused to sanction the flight. So, whether or not MacDonald crashed near Rockall will never be known.

The *Endeavour*

At eight thirty-five on the morning of March 13, 1928, Captain Walter Hinchliffe and his copilot took off from Cranwell, England, for Mitchell Field, Long Island. It was believed that his flying companion was Captain Gordon Sinclair, but this was a ruse. His copilot was in fact the Honorable Elsie Mackay, thirty-four, an attractive actress and aviatrix, and the daughter of Lord Inchcape, who ruled the P&O shipping empire. She was one of the richest heiresses in England. In her drive to be the first woman to fly across the Atlantic, she had persuaded the Air Ministry to put her in touch with Hinchliffe and swore them to secrecy. She knew if word of her intentions reached her father, he would stop at nothing to prevent the flight. She assured Hinchliffe ten thousand pounds in prize money and, in addition, insured his life for the same amount.

Hinchliffe, Mackay, and Sinclair arrived quietly at the airfield at dawn. Their mechanic waited beside the plane, the *Endeavour*, silhouetted against the morning light. It was a Stinson Detroiter with a forty-five-foot wingspan, a 225-horsepower Wright Whirlwind, and a five-hundred-gallon fuel capacity. Steady and reliable, it was a favorite with distance fliers.

As Hinchliffe and Mackay prepared to climb into the plane, Hinchliffe asked Sinclair to say nothing of his copilot's departure on the flight. Sinclair agreed, and also said he would disappear for a day or two to make the ruse complete. If the tall, striking airman had any doubt of reaching America with his companion, he never revealed it. Nor did Elsie Mackay, whose complete confidence in the one-eyed aviator remained steadfast. Hinchliffe had lost his eye to a German bullet during World War I, and wore a patch over it. It did not detract from his flying skill.

At eleven-thirty A.M., with the news services already humming, the *Endeavour* was seen passing over Kilmeadan, Ireland. A storm was raging over Cork at twelve-thirty when the Stinson was observed heading westward, battling the elements and heading into uglier weather. Observers were certain Hinchliffe and Mackay would turn back, but the plane pushed through the rough clouds safely and was again sighted at one-thirty that afternoon over the Irish coast at Mizen Head, about 465 miles from Cranwell. Later in the day it was sighted at sea by a French steamer as it passed low overhead. It was still maintaining a fair ground speed of ninety-three miles an hour, despite stiff headwinds. To all appearances, the *Endeavour* was off to a fine start.

The next day no ships reported seeing the Stinson, and at Mitchell Field the waiting crowd thinned out and drifted homeward. A few hopefuls lingered long into the night, scanned the eastern sky, and strained to hear the sound of an engine. Several false reports came in but only one, from Old Orchard, Maine, sounded plausible. At five after one Wednesday morning, the fourteenth, at a time when the *Endeavour* could have been approaching the Continent, the engineer of a passenger train looked out of his cab and saw the lights of a low-flying plane. If it was the British pilot and his companion, they could have crashed just offshore, lamentably close to their goal. The most prevalent theory was that head winds slowed the Stinson until its fuel was exhausted from repeated battles with Atlantic squalls and the plane was swallowed by the turbulent sea without a trace.

The saga of the *Endeavour* did not end here, however, for strange postscripts to the Hinchliffe-Mackay story began to occur.

Eighteen hours after the *Endeavour*'s departure, two old friends of Hinchliffe, RAF Squadron Leader Rivers Oldmeadow and Colonel G. L. P. Henderson were heading for England aboard the P&O liner *Barrabool*, then steaming south of the Canaries. Neither knew of Hinchliffe's flight. The sea was calm and Oldmeadow was asleep in his cabin. The door burst open suddenly and Henderson barged in, his face showing wild alarm.*

"Rivers," he said, "something ghastly has happened! Hinchliffe has just been in my cabin. Eye patch and all. He woke me up. He kept repeating over and over again, 'Hendy, what am I going to do? What am I going to do? I've got this woman with me, and I'm lost. I'm lost!' Then he disappeared in front of my eyes! Just disappeared."

Oldmeadow tried to calm his friend. He poured three fingers of straight scotch, and it helped. Henderson finally went back to his cabin, and tried to sleep.

Three days later, a news sheet was posted on the ship's bulletin board. It read: CAPTAIN RAYMOND HINCHLIFFE MISSING AFTER TRANS-ATLANTIC ATTEMPT.

Then, a month after the *Endeavour*'s disappearance, Hinchliffe's widow received a letter from an amateur psychic who said she had received several messages from Captain Hinchliffe. The spiritualist also informed Sir Arthur Conan Doyle, England's most respected spiritualist, of the messages. Doyle first investigated the authenticity of the woman's spirit communications through an experienced spirit medium, then wrote to Mrs. Hinchliffe that the communica-

---

*John Fuller, *The Airmen Who Would Not Die* (New York: G. P. Putnam's Sons, 1979).

tions appeared to be a true message from her husband. Shortly afterward, Mrs. Hinchliffe attended a seance and, through the medium, received a detailed account of the flight from a source that claimed to be her husband. He told how they had encountered bad weather nine hundred miles out—a terrific gale that damaged the left wing strut and tore the plane's fabric. They turned south toward the Azores to escape the storm but were forced lower and lower. Then the compass became erratic and one of the spark plugs failed. Hinchliffe said they made a successful ditching one mile north of the Azores at about three A.M. He said he tried to swim ashore, but after twenty minutes in the strong current he went under. Elsie MacKay's end, he said, was peaceful. She drowned while unconscious in the airplane.

In spirit sessions that followed, Mrs. Hinchliffe received information through the medium that only her husband could have known. She became convinced that her husband had spoken to her from beyond.

One thing remained, however, to puzzle those who had sought the answer to the two fliers' fate. Nine months after the *Endeavour*'s disappearance, a landing gear wheel was washed ashore at County Donegal in northern Ireland. It bore a serial number that linked it unquestionably to the *Endeavour*. If the plane had ditched in the Azores, why was its wreckage found in Ireland? The answer could only lie in the prevailing Atlantic currents, whose drift slowly floated the wheel ashore. But then, we shall never know beyond doubt.

The Secret Flight of the Mail-Route Pioneers

Unnoticed and unheralded, a graceful Bellanca floatplane lifted from Detroit's seaplane port on July 28, 1931. It swung to the northeast and settled on to a great-circle flight to Copenhagen. At the controls was Parker D. Cramer, an American World War I pilot and former member of the Sir Hubert Wilkins Antarctic expedition. Cramer had tried twice before to fly the Atlantic and had failed both times. In 1928, he and his companion Bert Hassell were forced to walk out over the Greenland ice cap after being forced down. In 1929, when the *Chicago Tribune* sponsored Untin Bower's Chicago-Berlin dash, Cramer was the copilot. The flight ended in the ocean and, luckily, both pilots were rescued.

Cramer's companion and copilot on this unannounced flight was Oliver Pacquette, a Canadian wireless operator. The well-planned air journey was not undertaken to break records or to establish a distance mark. Although it was still a challenge, the glory of the transatlantic dash was fast fading. Cramer and Pacquette were adventurers of a different sort, visionaries of a far-reaching dream. They were off to chart the first transatlantic air mail route on a course north of the Arctic Circle. It would connect the United States with the land of the Vikings.

For seven days, Cramer and Pacquette droned northward over the Canadian wilds. Except for one bad day at Hudson Bay the weather was good. On August third, they arrived at Hosteinborg on the west coast of Greenland. Thus far their flight had not attracted the attention of the news services; the pair smiled at the idea of having "put one over on the news boys."

The following day, in a brisk five-hour flight at ninety-six hundred feet, they crossed the Greenland ice cap en route to Angmagsalik, from where they planned to jump off for Iceland. As they soared high above the inland ice fields shimmering in the bright sunlight, Cramer thought the sun-illuminated snow fields were wonderful. "They'll be a great tourist attraction someday," he remarked to Pacquette. The most hazardous leg of their journey, or so they thought, was behind them.

News of the air venture hit the presses before they left Greenland. The full meaning of the flight was revealed by E. G. Thompson, president of the Thompson Aeronautical Corporation of Cleveland, who had selected the men to chart a proposed route for future air-mail pilots. "The route will be over the Greenland ice cap and the North Atlantic," Thompson told reporters. "Cramer's flight is the first ever undertaken and completed across the ice cap. Naturally, we're pleased with the success of their trip thus far, but we anticipate that much remains to be done in the way of study and preparation before we shall feel qualified to apply to the Post Office Department for a mail contract."

Thompson told the newsmen that another plane was being readied and would take off as soon as Cramer and Pacquette had finished their flight. It was to be followed by other flights on a monthly schedule. "Their purpose will be to determine just what we must face in all seasons," Thompson concluded. "We want to fly over these little-known regions during both freeze-up and break-up periods."

Encouragement for the air-mail flight came from Washington, where Assistant Postmaster-General Glover predicted the day was not far away when planes carrying mail to distant points would be a common occurrence.

Meanwhile, Cramer and Pacquette pushed on, winging their way toward Iceland. Almost as revolutionary as their journey was the fact that their aircraft was powered with a highly unusual engine—a diesel. This unconventional power plant was developed three years earlier by the Packard Motor Car Corporation and was, to say the least, unique. It was the first air-cooled diesel and the success of the flight with this type of power would not only be a notable first-time feat in itself, but would boost the small group of aviation diesel proponents. Later, in the thirties, the huge Dornier flying boats of Germany, with their liquid-cooled in-line diesels, would make regular South American runs, but thus far only America had produced a successful air-cooled radial.

The engine had many advantages. It consumed less fuel and allowed a greater range. It required no carburetor and no ignition system, thus eliminating the majority of engine troubles. The absence of the engine's electrical system permitted Pacquette's coded transmissions to be free of radio ignition interference caused by high-voltage magnetos. Fire hazards were practically nonexistent. The fuel oil could not be accidentally ignited; it would burn only when properly atomized by the high-pressure fuel pump. Above all, the engine seemed perfectly adapted to high-altitude, cold-weather flying. The fuel and air mixture would not preignite or detonate in the cylinders as it could in a gasoline engine, and carburetor ice—the dreaded hazard of North Atlantic flying—was nonexistent in an intake system that didn't have a carburetor. Except for the slight additional weight of the engine, which was designed to withstand greater combustion pressures, it seemed to be the ideal answer—a trouble-free, reliable powerplant.

In the early morning of August sixth, the Detroit-to-Denmark fliers approached the coast of Iceland. They were able to contact Reykjavik radio when only two hours out of Angmagsalik, at which time they requested a report on visibility and weather conditions along the northwest coast. Reykjavik informed them that their intended destination of Isafjordur was covered with fog. Cramer altered course for Reykjavik, two hundred miles farther south where the weather was clear. At three-twenty A.M., the men flew the Bellanca over the capital city at nine thousand feet without seeing it. A few minutes later they discovered their error, turned back, and landed the floatplane near the beach.

After they moved the monoplane into the harbor and supervised the refueling, the men slept. At two-ten on the afternoon of the seventh, they were off again, this time for the Faroes, a small cluster of islands 250 miles north of Scotland. They faced nine hundred miles of treacherous water over a route rarely traveled by ships, even in the summer season. Quick help to a downed plane in this region was out of the question.

The first hint of trouble was received at Thorshavn, the only town on the islands. Late in the afternoon a radio message crackled through the wireless station on the desolate, windswept outpost.

Forced to go down owing to engine trouble. Should like to obtain exact bearings. Can you assist, please?

Thorshavn could and did. After taking a bearing on the plane at sea, they also sent a dispatch to the London *Daily Mail*, informing them of the forced landing. Public concern for the daring men spread. When veteran fliers heard the news, they shook their heads. Cramer and Pacquette were doomed, they said, knowing full well the perils of coming down on the ocean in a small floatplane.

What caused the forced landing at sea was never known, but the remarkable feat that followed aroused the admiration of aviators on both sides of the Atlantic. After skillfully setting the Bellanca down on the choppy water, Cramer made the necessary adjustments, crawled back into the bobbing plane, and lifted it masterfully from the rough water. It was a narrow and dramatic escape from death.

When they reached Sydero Island in the Faroes that night the surveyors rested only long enough to refuel and to gain strength for the final two "easy stages" of their voyage. Early next morning, August eighth, they departed for Norway, with their destination either Bergen or Stavanger, depending on the weather.

Cramer and Pacquette were tired from their long hours in the air, but their spirits were lifted by the thought that this very night they would dine in Copenhagen. Unknown to them, the city was preparing a gala celebration during which they would be presented with the gold medal of the Aeronautical Society.

A few hours out of Sydero, the weather over the North Sea turned foul. An ominous fog bank closed over the ocean, and the Bellanca was buffeted by rough winds and squalls. Wisely, Cramer turned toward the Shetlands and landed at Lerwick to await clear weather. The next day conditions improved, and they decided to push on for Bergen.

When the graceful floatplane slipped easily into the sea spray off Lerwick and turned south toward Norway, it was the last time Cramer and Pacquette were seen. Somewhere east of the Orkney Islands something tragically ended the great dream of these men. Whether they had engine trouble, wing ice, or structural failure due to a sudden violent squall, was never known. The dots and dashes of Pacquette's telegraph key gave no hint. Only one thing is certain. The end was quick and violent; they had no time to tell their plight before they smashed into the icy waves of the North Sea.

Five weeks later, the British trawler *Lord Trent* came upon some drifting fragments of a plane wreckage, but no bodies. They fished one of the floats from the water and sent a wireless to London with the serial number of the float. The Associated Press relayed the data to New York, where K. D. Vasler of the Edo Aircraft Corporation consulted his files and announced, "This was the serial number of the floats on Cramer and Pacquette's Bellanca."

It was six months later, in March of 1932, that the final chapter was told. The crew of Dutch Trawler 130, plowing along at latitude 59.38 north, longitude 3.42 east, spotted a floating package in the water. It was not far from where the earlier wreckage had been seen. On opening the waterlogged package, they found Cramer's pilot's license, the permit for his transatlantic flight, the Bellanca's registration papers, and a letter from Cramer's mother. The contents were turned over to the American Consulate in Amsterdam.

Thus, the flight that began in secrecy ended in secrecy, and the North Sea keeps its secrets well. If Parker Cramer and Oliver Pacquette could look today on the great air-transport system of silver jets that carry mail and passengers over the very route they had blazed across the icy Arctic blue, they would say with pride, "We were first."

8
The Dole Derby

Back in 1927, aviation's wildest gamble was the twenty-four-hundred-mile dash to the Hawaiian Islands. When it was over, five aircraft were wrecked, three were lost at sea, and ten persons were dead. It proved to be little more than a mad dash for money and glory. And not a trace of the pilots and planes downed between California and Hawaii on the nights of August sixteenth and seventeenth, nineteenth and twentieth, was ever found.

A year earlier the Atlantic claimed the world's attention, for both 1926 and 1927 had been tragic years for ocean crossings. Most of the planes and pilots that plunged into its icy waters were never seen again. It was a time of financial inducements and promises of quick fame. Any pilot who could get his hands on an airplane and could muster the courage to head it eastward over the water, was in the running. It didn't matter whether he had had any overwater navigation experience. Most of the pilots did not comprehend the meaning of the word. Europe, they reasoned, was a big continent. So, if they could carry enough gas, hit the right weather, and fly long enough without going to sleep at the controls, they figured that sooner or later they would hit it.

At least thirty persons had crossed the Atlantic in airplanes and dirigibles before the Lone Eagle made the first solo crossing in May of 1927. To Lindbergh went the coveted twenty-five-thousand-dollar first prize offered by Raymond Orteig. In Paris, Ambassador Myron T. Herrick offered another twenty-five thousand dollars to the first person who would fly from Paris to his home city, Cleveland, Ohio. In Dallas, William E. Easterwood, Jr., offered a similar sum for the first Dallas-to-Hong Kong flight.

One other man of prominence was interested in a certain long-distance ocean attempt. He was forty-nine-year-old James Drummond Dole, the Hawaiian pineapple tycoon. While in California on business, he read of Lindbergh's success. Precisely what Jim Dole hoped to accomplish when he announced a twenty-five-thousand-dollar purse for a California-to-Hawaii dash is argued to this day. Was it a desire to speed air transportation to the islands? Or a publicity stunt to open Hawaii as a tourist land? Whatever his reason, his sponsorship of the transocean marathon, the greatest air race of its time, resulted in financial tragedies, frustration, and sudden death.

Twenty-five thousand dollars pitted against any reasonable hazard is still enough of a challenge to send today's air adventurers scrambling to the starting line. Back in 1927, it sent them wild. Even the ten thousand dollars second-place money sounded good. Dole bettered Orteig's offer.

From the beginning, several things went awry in the Dole Derby. The hazard was not reasonable nor did the prize money offset the risk to men and machines. True, the Pacific fliers only had to cover twenty-four hundred miles (compared with Lindbergh's thirty-six hundred), but Hawaii wasn't as easy to find as Europe. And for most of the trip, they would be bucking head winds. When it was all over, aviation authorities claimed the Dole Derby had set aviation back twenty years.

Attempts to reach Hawaii by air were not new. In 1925, Navy Commander John Rodgers tried it in a flying boat and was forced down three hundred miles short of his goal. After drifting for nine days, he and the crew of his disabled plane were sighted by a patrolling submarine and towed to the islands.

When Jim Dole notified the press of his offer four days

after Lindbergh landed in Paris, pilots from all over the United States flocked to Oakland to plunk down the one-hundred-dollar entrance fee. There were over forty of them—barnstormers, ex-war pilots, Hollywood stunt pilots, and air-mail pilots—all eager to cop the "easy money." They were unperturbed that Lloyd's of London did not think much of the venture. Normally, this insurance firm would insure almost anything, but now it refused to underwrite any of the Dole contenders.

In California, none of the Dole pilots was giving much thought to the overwater dangers. Several were inexperienced in long-range navigation over land, with all its identifying landmarks, and most of them had never been over the water and out of sight of land. A scant two or three had radio equipment and the knowledge of how to use it. And what a motley assortment of flying machines they brought to the starting line! One plane was so ridiculously slow it couldn't carry enough fuel to fly the distance. It nosed over on every landing, so Tex Lagrone wisely withdrew his Air King biplane, *City of Peoria*.

The breakneck rush to California and the rash eagerness of the entrants to tackle the Pacific on the early August starting date worried officials of the newly formed Aeronautics Branch. They feared that tragic consequences would result over the open water. They intervened and insisted on rigid examinations for both planes and pilots. This had some safety value since the shortsighted entrants who were long on courage and short on judgment were forced to drop out. By August eighth, only fifteen entrants remained. These men pulled numbers from a hat to determine their takeoff order, while James Dole, undismayed, happily commented, "It's shaping up to be a regular free-for-all." It certainly was.

At the very height of the preparations, four pilots stole some thunder from the Dole Derbyists. On June twenty-eighth, Lieutenants Maitland and Hegenberger of the Army became the first men to fly nonstop to Hawaii. Then, on July fourteenth, airmail pilot Ernie Smith and his navigator,

Emory Bronte, duplicated the feat. None of them were Dole contenders, and the racers who worked frantically at Oakland Airport could only salve their hurt pride with the philosophy, "It just proves it can be done." They overlooked the fact that Maitland and Hegenberger flew a trimotored Fokker, and Smith and Bronte, who barely reached the island, crash-landed into the treetops of Molokai when their fuel ran out. And both planes carried radios, the greatest single aid in transocean travel.

The race was plagued with misfortune from the beginning. In the hectic, tension-filled days before the starting gun sent them lumbering down Oakland's seven-thousand foot runway, three planes crashed during tests and three men were killed. Death struck first on August tenth. George Covell and Dick Waggener, United States naval officers, took off from San Diego in their Tremaine, the *Humming Bird,* to fly to Oakland for the start two days later. Their monoplane was heavily loaded with fuel and the engine was not operating well. Within minutes after takeoff, they crashed into a dune at Point Loma and burned to death. Covell had picked number thirteen from the hat.

The next day, the eleventh, Jim Griffin flew his huge and clumsy twin-engine triplane, the *Spirit of Los Angeles,* from Long Beach to Oakland. As he throttled back to begin the landing approach, spectators saw his glide was too flat. Suddenly, off the edge of the field and over the mud flats, the ungainly triplane suddenly stalled and flopped awkwardly into the swamps that bordered San Francisco Bay. Griffin and his two passengers were not injured, but the plane, sponsored by cowboy star Hoot Gibson was definitely out of the running.

August twelfth was set as the starting date for two reasons: On that day in 1898, Hawaii became a territory, and, too, a full moon was scheduled to rise at sunset—a moon that would light the fliers' way through the long night. This was the best possible night-flying condition that could be had at a time when instrument and radio navigation was still an undeveloped art. When the big day arrived, however, no one was ready. Several planes were still being modified, and government officials had not finished their tests of the pilots' fitness. Race officials reestablished the starting date as August sixteenth.

At Vail Field, outside Los Angeles, another entrant, the *Angel of Los Angeles,* underwent tests before leaving for Oakland. Arthur Rodgers took off in his tricky two-engine Bryant monoplane and began to curve around the field. The airplane, said to be capable of a top speed of 145 miles an hour, was favored by many to win. While still at a low altitude—125 feet—the plane started an uncontrollable slipping turn to the left. Rodgers struggled to bring it level, then, too late and too low, he abandoned the attempt. As his young wife Anna and designer Leland Bryant watched, he stepped into space and pulled the ripcord. The canopy of his parachute caught in one engine and part of the tail section and pulled him to his death as the plane plummeted to earth. The death toll now stood at three.

The *City of Oakland,* a Travelaire, withdrew. So did the *Miss Hollydale,* an International owned by Frank Clarke and Charlie Babb.

By noon of August sixteenth, only eight of the forty-odd original contenders were ready at the starting line. A crowd of fifty thousand, sparked by the publicity, crashes, and excitement of the past two weeks, lined the field to cheer them off. Everything was in readiness. Ten merchant vessels and eight destroyers were spread out along the length of the course. They were to blink identifying lights to the planes when they passed overhead and relay their positions to the mainland. The four planes with radio equipment planned to home in on a radio beam transmitted from Maui, but only one of these planes, the *Woolaroc,* had two-way communication with ship and shore. The latest weather data, pieced together from ships observing at sea, indicated the winds in this part of the Pacific would be moderate, northwesterly, and would swing about sharply to the east or southeast on the last lap, giving the racers a boost into Oahu.

The men climbed aboard their planes, and the engines of

Beech Aircraft Corporation

The *Woolaroc* Travel Air Model 5000. This five-place monoplane, piloted by Art Goebel and navigated by Lieutenant Williams V. Davis, was the winner of the 1927 Dole Derby.

the Dole racers popped, banged, and vibrated to life. Slowly, according to their takeoff position, they taxied in single file to the starting circle and awaited the starter's flag. Scheduled to be first off was Benny Griffin's and Al Haney's blue-and-yellow Travelaire, *Oklahoma*. It paused inside the starting circle, its engine ticking over impatiently, and as the starting flag flashed down, roared to sudden life. The heavily laden plane surged forward, slowly gained speed, and staggered cumbersomely into the air at two minutes past twelve. The race was on!

One minute later the starter's flag dropped again, and Goddard and Hawkins in their silver, high-wing, open-cockpit monoplane, *El Encanto,* started down the runway. As it gained speed, it began to swerve uncontrollably right and left. In a flash, it veered sharply to the right in a vicious ground loop. The landing gear collapsed and the plane skidded to a stop in a cloud of sand and dust. Neither pilot was injured, though both were emotionally shaken by their bad luck. The plane was a total washout and, fortunately, the gas tanks did not burn or explode.

At eight past twelve, the *Pabco Pacific Flyer,* a Breese monoplane with Livingston Irving at the controls, got the signal. With its tremendous load of 450 gallons of gasoline, it inched forward sluggishly. Then it began to move off the runway, and Irving, unable to control it, cut the switch and rolled to a stop. A tractor was used to pull the plane back to the starting circle.

Next, the *Golden Eagle,* a hundred-mile-per-hour Lockheed Vega flown by Jack Frost and Gordon Scott, took easily into the air and roared past the wreckage of the *El Encanto* and over the *Pabco Pacific Flyer* being returned for a second takeoff attempt. The *Golden Eagle* was highly favored to win. It was without a doubt the fastest plane in the race, with the best instrumentation. One minute later, Augy Pedlar, Lieutenant Silas Knope, and their passenger, Mildred Doran, a pretty twenty-two-year-old Caro, Michigan, schoolteacher, roared skyward in the *Miss Doran,* a Buhl sesquiplane. Mildred Doran was the only woman in the race. The *Aloha,* another Breese monoplane, piloted by Martin Jensen with ocean captain Paul Schluter as navigator, lifted easily and climbed westward. The seventh plane off was the *Woolaroc,* a Travelaire flown by Arthur Goebel and Bill Davis, and bringing up the rear were Bill Erwin and Alvin Eichwaldt in their green-and-silver *Dallas Spirit.* It later came to light that these two had dreams far beyond those of the other Dole fliers. They were out to capture two prizes, the Dole purse and the additional twenty-five thousand dollars for the first Dallas-to-Hong Kong flight. Added to these prizes was William E. Easterwood's promise of his personal check for ten thousand dollars which would bring the total to sixty thousand dollars. If ever two men had reason to set a distance record, this was it. After their Swallow monoplane became airborne, the huge crowd began to relax and drift homeward. Each spectator wondered which plane was going to reach the islands first.

Then someone pointed over the bay. Coming in low, streaming a white-silver streak of raw gasoline, was the *Miss Doran*. Pedlar was dumping fuel overboard to lighten the plane for a landing. He touched down and taxied hurriedly to the line. "It's backfiring and missing!" Pedlar yelled over the engine noise. He could have saved his breath; the engine sounded like a bucket of bolts to the most inexperienced onlooker. Pedlar shut down the popping machine and his mechanics scrambled over the cowling. The red-white-and-blue Buhl had managed to struggle to eight hundred feet before a gasoline supply line became blocked.

The crowd shouted again. A second plane, the *Dallas Spirit*, was coming back. Even before it landed, the trouble was obvious—a large fabric tear along the side of the fuselage. Much of the covering was ripped away and shreds were whipped briskly in the slipstream. The damage would take hours to repair. Erwin and Eichwaldt were out of the race. The *Dallas Spirit* had barely come to a standstill when the *Oklahoma* roared low over the crowd, its engine popping and trailing black smoke. As it landed and turned in beside the other two, Griffin's mechanics shook their heads. It was all over for the *Oklahoma*, too. Its overworked engine had blown five cylinders and was a total loss.

While mechanics worked frantically to change plugs on the *Miss Doran*'s hot engine, Livingston Irving climbed back into his *Pabco Pacific Flyer* and tried for a second takeoff. He was already an hour and twenty minutes late, but he believed his chances were still good. Halfway down the runway he lifted the monoplane briefly. Then it settled on one wheel, bounced, lifted painfully, and settled again. It started into a wide, skidding ground loop. Finally, it toppled over on its side in a cloud of dust. The crowd gasped and tensed for an explosion. It never came, and Irving, physically unhurt but crushed in spirit, clambered out as screeching fire trucks and ambulances pulled to a stop beside him. His fellow pilots quickly drove his wife, carrying their small daughter in her arms, to the wreck. A downcast Irving greeted her with, "Well, my dear, I won't get to Honolulu now."

At the opposite end of the field, pretty Mildred Doran, shaken and pale over the aborted start of the Buhl, was nevertheless determined to go on. Several in the crowd shouted out demands that she be removed from the race. But she smiled faintly and climbed back aboard the Buhl as the mechanics snapped the final piece of cowling in place. Shortly after two P.M., the trio lifted off again and headed into the gusty mists beyond Fort Point and the Golden Gate. It was not until some time later that one of the mechanics thought to ask whether the *Miss Doran*'s tanks had been refilled to replenish what Pedlar had dumped. They had not, and many of those who sat up all that night to ponder the progress of the fliers wondered if this little oversight was really important. No one would ever know for certain.

Now began the vigil. Those who waited throughout the night in the darkened airport hangars discussed the probable hazards awaiting the pilots. Darkness would overtake them at about eight P.M., then would begin the long night. Every pilot carried life belts and a life raft, as well as emergency food supplies. Jack Frost's Vega, the *Golden Eagle*, had cork slabs fitted into the wings, a sealed fuselage, and compressed-air cylinders to inflate flotation bags. It alone could remain afloat for days if forced down. The big worry, they all agreed, was the weather. A spin or a spiral dive while flying blind would be fatal.

One hour and twenty-eight minutes out of Oakland, the *Aloha* passed over the SS *Silver Fir*. Working a time-distance problem, navigator Schluter established their speed at a steady eighty miles per hour. A few minutes later the radio operator aboard the vessel heard another engine above the clouds and tapped out an inquiry. The *Woolaroc* replied, and as daylight faded, Bill Davis began to send out hourly position reports while simultaneously homing in on the directional beam transmitted from Maui. At eight o'clock, the SS *Wilhelmina*, five hundred miles out of San Francisco, picked up a strong signal from the *Woolaroc*'s key as it passed nearby.

After sunset, problems began to develop rapidly for the aviators. Things went somewhat better for the crew of the *Woolaroc* than for the *Aloha*'s team. Arthur Goebel was more experienced in instrument flight than was Martin Jensen, who had difficulty finding his way blindly through the night and the thick fog. At one time during the night, as he groped through the inky blackness, Jensen felt the underside of the *Aloha* slap into a whitecap. He fought off panic, pulled up, and gave the engine full power. When he leveled off he was in a cold sweat. The night ticked slowly on with no reference points, and the plane fell into several spins. Each time Jensen recovered before the churning waves could reach them. It was during this night-long struggle to keep the *Aloha* reasonably level and on course that Schluter passed him a note. "Lost until I can get a fix on Polaris."

At two o'clock Wednesday morning, the SS *Manulani* heard the distant drone of a plane engine to the southwest. Moments later, another. The ship's radio operator raised one, the *Woolaroc,* and passed the word to the mainland. But no one had any inkling of who was flying the second plane or the whereabouts of the remaining two still unaccounted for.

Meanwhile, Honolulu was wide awake. Long before dawn thirty thousand spectators jammed the road to Wheeler Field. Only the position of the *Woolaroc,* which had transmitted faithfully every hour, was known. It was dead on the directional beam, and at six A.M. was reported six hundred miles out. But where were the others? The *Miss Doran?* The *Golden Eagle?* The *Aloha?*

Just before noon the *Woolaroc* rasied land. Manuai! A short time later Molokai and Diamond Head slipped under its wings and a military escort of planes roared out to meet it. At twelve-twenty, Goebel and Davis landed to the thunderous roar of the crowd, first-place winners. Their time: twenty-six hours and seventeen minutes. Now the big question was who would follow them.

During the black hours over the Pacific, Martin Jensen struggled to keep the *Aloha* on course. Vertigo and sensory illusions confused him. Below him was inky blackness; above, a solid cloud layer. The only thing he knew for certain was that they were lost somewhere between California and Hawaii, and as the night wore on, their hopes grew darker. Then, the nightmare broke into a beautiful dawn. Schluter, who had been unable to get a star fix once during the night, was certain from the record of his elapsed time that they were on a line (north or south) of the islands at eight-thirty. At nine o'clock Jensen passed a note to Schluter. "Where from here?"

Schluter scribbled on a pad and passed the answer forward. "Circle till noon."

Until the sun was at its zenith, Schluter, who was accustomed to surface ship navigation, couldn't get a celestial fix. For the next three hours the pair flew in wide circles as their gas began to run dangerously low. One after another the tanks ran dry. At noon, Schluter quickly made a sun shot. His report: two hundred miles north of Hawaii!

Leaning out every ounce of gasoline, Jensen reached out for Diamond Head. After what seemed an interminable time, they spotted the island, hopped over Koolau Range, and let down for Wheeler Field. They landed with four gallons of fuel remaining, twenty-eight hours and sixteen minutes out of Oakland, as second-place winners.

By late afternoon, hope was fading for the remaining two planes. At four-twenty, Frost's *Golden Eagle* would have exhausted its fuel; at eight-twenty that evening, the last tank of the *Miss Doran* was dry. No one knew when or where their flight was stilled or what fate befell them over the Pacific on that long night of overcast, swirling fog and darkness. Somewhere in that empty sky they met trouble, and somewhere in that empty sea they crashed. And the whitecaps rolled over them and smoothed away all trace of their graves.

The next day, the eighteenth, Jensen and Goebel searched

with Army and Navy planes for signs of wreckage. Reports of a signal flare on Mount Mauna Kea came in during the night, but it turned out to be a rancher testing a new type of gasoline lantern. Then a red-white-and-blue object was reported afloat in the water east of Maui. Hopes arose that it was the *Miss Doran*. It wasn't; it was only a sampan.

Acting Secretary of the Navy Eberle authorized a sea search for the missing planes. It covered 540,000 square miles and lasted a week. The seven destroyers, submarine tenders, cruisers, and aircraft carriers—forty Navy vessels in all—comprised, up to that time, the largest fleet to be used in a search in naval history. Jim Dole personally put up a reward of twenty-thousand dollars for the finding of either plane, Bill Mollaska offered ten thousand dollars for the *Miss Doran,* and George Hearst, Jr., put up another ten thousand dollars for the *Golden Eagle*. No one collected.

Death was to touch its icy fingers on two more daredevils before the full account of the Pacific marathon would be closed. On Friday morning, August nineteenth, Bill Erwin flew his repaired *Dallas Spirit* around the bay to test his two-way radio. It had been given the call letters KGGA and was installed while the repairs were being made to the Swallow's fuselage. When Erwin landed he was in excellent spirits. He reported everything fine and ordered his fuel tanks topped to their limit of 450 gallons.

Bill Erwin was no beginner to flying. He learned to fly in 1917 with the Signal Corps, went to France with the 1st Aero Squadron, and rose to the rank of captain. Flying Spad XIs, he shot down eight enemy planes, was decorated, and awarded the Distinguished Service Cross. "Lone Star" Bill and his navigator, Alvin Eichwaldt, were thwarted in their try for the Dole prize, but now they declared themselves in the running for the Easterwood prize. Though they claimed they were off to search for the *Miss Doran* and the *Golden Eagle,* everyone knew otherwise. Hong Kong was their destination.

Five thousand spectators turned out to see them off. And

Dallas Morning News

Captain and Mrs. William P. Erwin in the *Dallas Spirit* following the unveiling ceremony at Love Field, Dallas, on August 6, 1927.

although neither the Navy Department nor the National Aeronautics Association approved of their flight, the two would-be record makers would not be turned aside.

Ted Dealy, who later became the publisher of the *Dallas Morning News,* was a reporter at the time and a personal friend of Erwin. In 1965, looking back on that fateful date, he said, "Erwin was thirty-one years old, the son of a Presbyterian preacher. He had no premonition he would lose his life in the Pacific. Before he took off he left a written message for his mother and a second letter for his sponsors. Each one evidences his deep religious beliefs and his confidence in the outcome of his flight. In the letter to his sponsors, Erwin wrote:

> Tomorrow begins the great adventure. Flying personifies the spirit of man. Our bodies are bound to the earth; our spirits are bound by God alone, and it is my firm

belief that God will guide the course of the *Dallas Spirit* tomorrow over the shortest route from the Golden Gate to the Isle of Oahu. . . . If we succeed, it will be glorious.

If we fail, it will not be in vain, for a worthy attempt could never result in a mean failure.

I believe with my whole heart that we will make it. I believed it when I first conceived it, and I believe it more strongly now. We will win because *Dallas Spirit* always wins.

But if it be His will that we should not make it, and from the exploration of the Pacific we should suddenly be called upon to chart our course over the Great Ocean of Eternity, then be of good cheer. I hold life dear, but I do not fear death. It is the last and most wonderful adventure of life. If something should happen to me I know I don't have to ask you to look after Mrs. Erwin. It broke her heart that she could not accompany me. She is my life, gentlemen, and the sweetest, finest, truest girl the Almighty ever created.

Knowing that she is safe gives me confidence and vigor for the trial. We will make it because we must, but whatever comes, I am the master of my fate, and God willing, the Captain of my soul.

"The Mrs. Erwin he mentioned was his wife Constance Ohl Erwin. She wanted to accompany her husband on the flight, but was ruled out because she was under twenty years of age. It was just as well, for the Erwins were expecting a child.

"When Captain Erwin left Oakland Airport on his last voyage into the air," Dealy concluded, "he had a small Bible in the right-hand pocket of his sack coat. He didn't mention it, but I saw it there as I was helping him into his flying suit."

Shortly after two in the afternoon, Erwin and Eichwaldt made an uneventful takeoff and headed out the Golden Gate. Twenty minutes later they began to transmit code messages. At three forty-nine, they reported the ceiling at seven hundred feet; they were flying at five hundred. At five-ten, they passed over the SS *Mana,* and at six fifty-eight, they reported their position at 35.30 north latitude and 130 west longitude—about 450 miles out. At seven-twelve the key crackled, "We have thirty miles visibility and are flying at nine hundred feet. Have seen nothing [of the *Miss Doran* or the *Golden Eagle*]."

At eight o'clock: "It is beginning to get dark."

Eight fifty-one: "SOS. [pause] Belay that. We were in a spin, but came out of it okay. We were sure scared. It was sure a close call. The lights on the instrument panel went out and it was so dark Bill could not see the wings."

Two past nine: "We are in an . . ."

Silence.

As the half-finished message broke off, radio operators on ship and shore waited anxiously. Several tried to tap out a message to KGGA. "Come in. . . . Come in KGGA. . . . Do you hear us, KGGA?"

KGGA did not hear. Apparently, Eichwaldt was trying to transmit, "We are in another spin," and never finished it. And so the *Dallas Spirit* changed its course and headed into that "Great Ocean of Eternity" that Bill Erwin had mentioned.

At dawn, ships converged on the place where the *Dallas Spirit* made its final call. They searched the area for a week without finding a sign of anything—not a trace of wreckage, not a scrap or fragment of anything resembling an airplane. Like the others, the Pacific had swallowed the aircraft completely. The flight paths of the three missing planes were over the most traveled sea lanes between the islands, crossed almost constantly by passenger liners, freighters, tankers, and naval vessels, yet not a scrap of floating debris from any of the aircraft was ever found.

In the weeks that followed, the public protested and critics blasted the Dole Derby as a blow to aviation's progress. James Dole came in for a major share of the blame, accused of thinking only of publicity for his pineapple empire. Dole admitted he felt an indirect responsibility. Newspaper and magazine editorialists who a few weeks before had praised his gesture as a milestone in aviation pioneering, now decried just as loudly his thoughtless encouragement of stunt flying.

Perhaps everyone was partly right, but the fact remains that, by hurrying to meet the starting deadline, many aviators who normally practiced safety in flight did dangerous and foolhardy things. In taking the short cuts they normally shunned, air safety took a back seat with several of the contenders. That it should have been riding as copilot was realized too late. Five planes were wrecked, three were lost at sea, and ten persons were killed—Waggener, Covell, Frost, Scott, Rodgers, Erwin, Eichwaldt, Pedlar, Knope, and young Mildred Doran, the first woman to be lost in a Pacific air crossing.

At noon on September 16, 750 miles out of San Francisco, the SS *Maui* cut its engines and drifted to a stop. The passengers and crew quietly assembled on deck and with heads bowed, listened as someone read from the Scriptures, "The Lord is my Shepherd . . . I shall not want. . . ." Then the voice of another, a member of the ship's crew, drifted across the Pacific waters. "Going home, going home. . . . I'm going home . . ."

The engines rumbled to a slow idle as the steamer made a huge circle. A shower of wreaths and floral pieces were dropped over the side. One of them, in the form of a Holy Bible, carried a message from the sixth-graders of Caro, Michigan. They had saved their pennies and nickles to send this parting prayer to those daring fliers who had gambled and lost all, and to Mildred Doran, their beloved teacher.

It said, "God Bless You Every One."

9
South to Rio

For more than five decades the mystery of Paul Redfern's disappearance was as puzzling as the day in 1927 when the slim, handsome young adventurer climbed into the cabin of his monoplane and flew into—what?

Years after the green-and-gold Stinson Detroiter *Port of Brunswick* vanished somewhere over South America, newspapers kept the story alive. Journalists were asking as late as 1961: Did Paul Redfern find his Shangri-la in the uninhabited wilds of British Guiana? Is he living as a white god somewhere along the snakelike Amazon? Someday, will he step out of the South American jungles as he allegedly said he would?

It is unlikely. Over a dozen searches have penetrated the dense brush of the Amazon Valley in the hope of finding a clue to the whereabouts of the missing pilot. Not one expedition has been even remotely successful. Not one scrap of evidence has ever been uncovered to indicate that Redfern survived his unknown rendezvous with destiny.

The mystery, then, is not whether Paul Redfern is alive, but where his green-and-gold monoplane disappeared and why.

The Roaring Twenties were in their seventh turbulent year. This was "the year the world went wild." Surely it seemed so. Lindbergh soloed the Atlantic; forty-odd contenders entered the Dole Derby. Anyone crazy enough to put up fifty-eight dollars could fly from Los Angeles to Seattle on a Pacific Air Transport plane. Hundreds of young eagles with dreams of adventure were planning and doing the feats that would shape aviation history.

A twenty-five-year-old former music student who had turned flier was among them. Born in Rochester, New York, in 1902, Paul Redfern was too young for World War I, but not for the postwar Jennies that barnstormed the country. Before he operated his own airport in Toledo in 1925, he had flown sixty thousand miles in forty states.

In 1924, Redfern was a pilot for J. R. Reichard Cigar Company of York, Pennsylvania. He spent some time in Toledo, Ohio, on an advertising campaign directed by C. C. Hillabrand, the company's Toledo distributor. In one of the first attempts at aerial advertising, Redfern flew over the city and dropped miniature parachutes with sample cigars attached.

Paul met and married Hillabrand's attractive auburn-haired daughter Gertrude in January 1925. After a honeymoon combined with an advertising campaign, they returned to Toledo where Paul leased some land and operated his own airport.

This lasted only a few months, for Redfern was restless. He had taken a position as a U.S. government "dry sky scout" and was stationed at the famed old Customs House in Savannah. His job was to locate stills from the air, signal their locations to ground parties, and to chase rumrunners at sea.

Redfern's young wife noticed her husband's restlessness. "Paul was determined, romantic, and adventuresome," she reflects today. "In his early flying days many newspapermen referred to him as 'Daredevil Redfern.' "

In 1926, Redfern outlined plans to win the Dole purse but changed his mind in favor of a more difficult goal, a nonstop solo flight from Brunswick, Georgia, to Rio de Janeiro. Early in 1927, on a leave of absence from his job with the Customs Office, he turned his efforts toward convincing several prominent Brunswick businessmen that the next air route to be conquered was one linking North and South America.

They agreed to back the venture and underwrote twenty-five thousand dollars for the plane and equipment.

Paul selected a Stinson Detroiter for the long-distance attempt. Although it was a single-engine aircraft, it had a high-lift wing, could fly at a moderate airspeed, and was equipped with the finest engine of its day, the proved 225-horsepower Wright Whirlwind—the same model engine that powered Lindbergh's *Spirit of St. Louis*. Special fuel tanks installed at the Stinson factory could be emptied in forty-five seconds if it suddenly became necessary to lighten the load.

In midsummer, Paul flew his Stinson from Detroit to Glynn Isle Beach, Georgia, to complete preparations for his solo flight. In a brief ceremony just prior to his departure, Mrs. Eugene A. Lewis, wife of a prominent Detroit business leader, christened the plane *Port of Brunswick,* in honor of the Brunswick, Georgia, businessmen who backed the flight. The plane bore the registration number NX773 on its wings.

Early in August, Redfern and his assistants began final preparations. J. J. B. Fulenwider was Redfern's nautical advisor; Paul J. Varner, chairman of the flight committee and one of the backers, assisted in selecting the equipment to be carried; and Captain D. M. Scaritt of the U. S. Department of Commerce was the official starter.

Redfern's survival equipment was more than adequate, provided he could set the airplane down in a clear area without injury to himself. He carried a small, inflatable raft, in case he was forced down at sea, emergency rations for ten days, fishing tackle, a collapsible rifle with one hundred rounds of ammunition, quinine and other medicines, hunting knives, matches, flares, a small distilling outfit, boots, and a parachute, one item most of the other distance fliers omitted in order to save space and weight. Redfern made no effort to conceal his belief that his greatest danger would be a crash landing in the unexplored wilds of the Amazon.

At the Stinson factory, Redfern had the Detroiter modified to carry 550 gallons of gasoline, enough to reach Rio if an

Photograph by Marion Johnson

Paul Redfern and his Stinson Detroiter *Port of Brunswick,* in which he attempted a nonstop flight to Rio de Janeiro.

average ground speed of ninety-two miles an hour could be held. It did not allow much of a reserve. After two days and two nights aloft, Redfern planned to land at Camp Affenso near Campo, outside Rio. He knew a fuel shortage or bad weather would seriously cut his flying distance, so he wisely selected Pernambuco on the easternmost tip of Brazil as his alternate destination. A fact frequently overlooked in studying Redfern's trip is that the intrepid young man's carefully planned flight would have outdistanced all previous ones.

The forty-six-hundred-mile span would, if successful, outdistance Lindbergh's flight of three months earlier by one thousand miles.

Three trial flights were made to confirm the airplane's balance, load-carrying ability, and airspeed. When Redfern finished, he was satisfied the plane would do the job, as were government officials and the Stinson representatives who carefully cross-checked every mechanical and aerodynamic detail.

When Gertrude Redfern received a wire from her husband that preparations for the flight were almost completed and that he was "rarin' to go," she left Toledo immediately for Brunswick to be with Paul for a few days prior to the takeoff scheduled for Saturday, August twentieth. Having flown with Paul many times, she was fully confident of his flying ability.

When news came of the lost Dole contenders and of Erwin and Eichwaldt's disappearance in the Pacific, it failed to shake Redfern's confidence. He said little about it and busied himself with last-minute details on the *Port of Brunswick*.

The departure date was changed several times. A tropical storm was reported to be brewing to the south, in line with the flight path. Impatient, Redfern pushed plans for an early Monday morning takeoff. But the weather grew worse and by Tuesday the horizon was dark and gloomy. On Wednesday a hurricane swished up the Atlantic seaboard and sent its skirting waves almost to the hangar doors. Everything was battened down tight.

Thursday saw a welcome improvement in the weather and Redfern made a final check of his route. The line drawn across his air charts to show the flight path cut across the extreme tip of Puerto Rico, passed directly over Grenada and Tobago islands. He hoped to make his South American landfall at St. Andrew's Point, British Guiana, then fly over Dutch Guiana and the northeastern part of Brazil until he reached Macapa on the north bank of the Amazon estuary. From here, it was thirteen hundred miles to Pernambuco on the east coast (his alternate), and 1,650 to Rio. Over Macapa, Redfern was scheduled to drop a colored flare—red if his gas was running low and he planned to land at Pernambuco, green if everything was satisfactory and he intended to proceed to Rio.

If the possibility of a forced landing in the jungle haunted Redfern, he failed to show it. In one of his final interviews he told the press, "Don't lose hope of my return for at least six months or more. If I should be forced down in the Amazon Valley, I believe I can live for months with the equipment I'm carrying."

On Friday morning, the dark-haired 138-pound aviator announced that "barring another hurricane, and God willing, I'm going to Brazil." Officials shook their heads; the weather picture was still doubtful in the Caribbean and over the Atlantic. He sent a last-minute telegram to his mother, Mrs. Blanche Redfern, who waited in Rochester. The message read, "Goodbye Mother dear. Will cable you from Brazil. Heaps of love."

Shortly after noon—a gloomy overcast midday—Redfern climbed into the cramped cockpit of his monoplane. His wife slipped into the cabin beside him. They talked for a few brief moments. Paul Redfern's final words to his wife were, "Goodbye, my darling, and don't worry—I'll see you real soon." (Gertrude was to meet him in Rio or wherever he might land, if it was necessary to change course due to weather or low fuel.) There was a lingering last-minute kiss and then the young girl stepped unhesitatingly to the ground and walked away from the plane. One of the flight-committee members handed Redfern a sealed packet containing letters to the president of Brazil and the mayor of Rio from the governor of Georgia and the mayor of Brunswick.

It was low tide. The waters had floated back from the marshes of Glynn and it was time to go. Redfern opened the throttle and began a slow, lumbering run down the long, eight-mile Sea Island Beach. As the heavily loaded plane, fifty-five hundred pounds in all, struggled to become airborne, a stiff crosswind from the sea threatened to flip it on its side. Redfern cut the engine, rolled to a quick stop, turned, and taxied back to the hangar. The crowd of three thousand wondered if he would call the flight off.

Paul Varner ran to the airplane and talked with Redfern. "I couldn't take the chance, Paul," Redfern shouted above the engine noise. "When I lifted the tail, the wind pushed me toward the water. Pull me higher onto the beach. I'll get off next time."

He did. A quarter of an hour later, with the Wright Whirlwind straining full out and the airplane now quartered into the wind, Redfern thrust it forward again and slipped heavily into the air amid the shouts of the crowd. It was twelve forty-six. Paul Redfern, Brunswick's adopted son, was on his way south to Rio.

News of the successful takeoff was flashed to Rio, where excited Brazilians were already preparing a gala welcome for the "Lindbergh of South America." Although Redfern was not due to arrive until late Sunday, the streets were already humming with emotional Latins anxious for the celebration to begin. At Campo, the United States Navy was ready with powerful searchlights to guide Redfern in if he arrived at night. Naval officers thought it would be even better if they could use their radio transmitters to direct him, but the *Port of Brunswick* carried no radio.

After the tiny speck disappeared low over the southern horizon, its position was reported several times. Shortly before midnight, the monoplane was sighted overhead at two thousand feet by a steamer three hundred miles east of the Bahamas. It was droning steadily along in a southerly

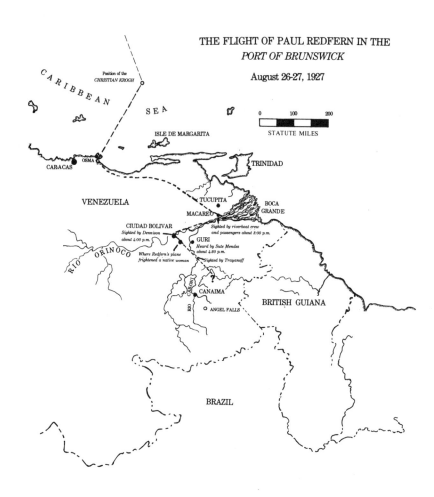

direction. The news was radioed to Station 4AQF in Nassau, which relayed the report to Brunswick by way of a radio operator in St. Petersburg, Florida.

In Brunswick, Redfern's wife and friends maintained a tense vigil. At Rio, Brazilians scanned the hourly bulletins while crowds began to congregate at Campo in anticipation of their hero's arrival. But even as Redfern winged over the Caribbean, armchair prophets were gloomily predicting his failure. One pointed out that the winds were sure to be unfavorable; naval navigators and hydrographers predicted he would miss his mark by 450 miles. They contended he would have to fly a total of twenty-four hours at night with no moon and that he would not have enough gas for what they were sure would be a sixty-two-hour flight.

On Saturday morning, ship's master Captain I. A. Hamre watched from the bridge of his Norwegian merchantman, *Christian Krogh,* as a lone plane, flying low, appeared over the northern horizon. As it circled his vessel, the crew could clearly make out the bold black letters on the fuselage: BRUNSWICK TO BRAZIL.

Three times the plane swooped low overhead as the pilot dropped messages in waterproof containers through the cabin window. The first one disappeared in the waves. As the second note fell, fireman W. T. Nodtvedt dived into the churning water and grasped the container. The message asked the captain to point his ship toward the nearest land. Hamre complied and swung the prow toward Venezuela, 165 miles away. Redfern zoomed low again and dropped a third note which was recovered. It asked the crew to "blow the ship's whistle once for each 100 miles—Redfern, Thanks."

As the thunderous steam bellowed from the ship's whistle, the plane lined up with the *Christian Krogh*'s path, wiggled its wings in thanks, and disappeared toward the Venezuelan coastline. Later that afternoon, excited natives watched the plane pass over the village of Tucupita, just north of the Orinoco Delta. Still later, it was seen crossing over the Canamacoreoa River, headed for Boca Grande. A heavy storm broke over the village of Macareito in the Orinoco Delta, but according to reports it occurred an hour before Redfern passed overhead, apparently unscathed.

This was the last time the plane was ever seen by white men—heading into the green nightmare of British Guiana and the Amazon River Valley.

As the hours moved up to the projected time of Redfern's arrival in Rio de Janeiro, Gertrude Redfern and Dr. Frederick C. Redfern, Paul's father, gathered with anxious members of the flight committee to hear the message flashed from Rio that the aviator had landed safely.

Long minutes slipped into even longer hours. There were no reports at all on Sunday, no word of any sighting along the proposed flight path. He had not flown over Macapa as planned. That evening Paul Varner solemnly announced to the press, "At four-thirty this afternoon we were certain that Paul Redfern was no longer in the air. His fuel would have been exhausted at that time."

Redfern had been forced down somewhere in the South American jungles. But where? Where in the hundreds of thousands of square miles of the wild continent? Did he crash in the Pacaraima Mountains of British Guiana? The Orinoco? The Amazon Basin? Or had he drifted westward, pushed by the prevailing easterlies, until he ran out of fuel in the unexplored heartland of central Brazil?

At Georgetown, British Guiana, the governor ordered a seaplane to patrol the jungles. A few days later, in an interview at the Great Northern Hotel in New York City, a commercial pilot, Robert Moore, claimed Redfern "was forced down before he flew five hundred miles from Brunswick. He was struck by a waterspout." Moore claimed to have seen Redfern dodging waterspouts in the Bahamas near Great Abaco Island on Friday afternoon. Moore's theory was proved wrong when the messages dropped to the *Christian Krogh* were received in Brunswick and positively identified as having been written by Redfern.

The first expedition for the missing aviator got under way

in September when Redfern's uncle, Richard Redfern, joined a search party headed by Richard O. Marsh of Rochester. Marsh, using two planes, started his search in the delta of the Orinoco and worked south. After several grueling weeks, they returned without finding a trace of the missing plane. The following June (1928), the citizens of Brunswick dedicated Redfern Field, in honor of their missing adopted son and his ambitious flight.

Between 1928 and 1937, expedition after expedition probed the South American jungles, tracking down every lead that could be found. None of them led anywhere. Art Williams, the naval pilot who had taught Redfern to fly, made extended flights over the route while he was working in South America, but like the others, his efforts were in vain.

In the years since Redfern's disappearance, thousands of words have poured from the pens of imaginative newsmen and feature writers. The romantics of aviation had a heyday. Some pictured Redfern as a crippled white god among the savages; others as a dazed, white-haired man stumbling aimlessly through the steaming jungles. There has not been one fragment of fact to substantiate these weird tales or any evidence whatsoever that Redfern even survived the crash of his Stinson.

The final and most elaborate expedition was arranged in 1937 by Colonel Theodore Waldeck, a friend of the Redfern family. One member of the party, Dr. Frederick J. Fox, died on the trip, and a second member, William Astor Chandler, almost lost his life. Colonel Waldeck eventually emerged from the jungles of British Guiana "convinced that Redfern crashed and died on the continent."

Until his death in 1941, the aviator's father steadfastly clung to the hope that his son might still be alive. But Paul's sister, Mrs. Ruth Sanders of Sumter, doubts that this could be. In recent years she has said, "In spite of the many rumors ... that Paul is alive, I don't believe it's possible that he could be after so many years without tangible evidence."

The last flurry of journalistic interest in the Redfern

mystery occurred in December of 1939. A Uruguayan seaman, Gustavo Adolpho Lentulus, who was temporarily stranded in Bayonne, New Jersey, told reporters this story. Clearly it is at variance with Colonel Waldeck's findings.

Shortly before he sailed from Argentina a few months earlier, Lentulus met a recluse, a Swiss resident of Buenos Aires, by the name of Hans Seigerst. He told Lentulus about a four-year expedition that he and three other men made into the jungles in 1932. According to Lentulus, Seigerst showed the sailor certain articles salvaged from the wreckage of a plane they found eight hundred miles from Buenos Aires. The items included a wristwatch, a buckle, and some papers, all taken from the remains of the pilot, who was still wearing a leather jacket. The papers, Lentulus claimed, were personally examined by him, and he identified a note that contained the abbreviation "U.S.A." and the word "Cal." It was signed "Redfern."

Seigerst was the only man to return alive from the expedition.

In 1969, renewed efforts to solve the mystery of Paul Redfern's disappearance—despite a trail grown cold in forty-two years—were undertaken once again. Two former pilots, guests at the Redfern Memorial dedication at Sea Island, met during the commemoration of the flier's distance attempt. They agreed to share their information and work together to solve the puzzle. Fred Lowen and George Bell (fictitious names) decided there might, one day, be yet another expedition for the lost adventurer—expedition fourteen—the final one.

Lowen, a formidable, square-jawed man with a ruddy complexion, is a decorated, former military pilot and one-time airline executive with an intense interest in aviation's past. For more than twenty years he has kept up to date on every major recovery effort of lost military and civilian airplanes, and has participated in successful recoveries. Bell, bespectacled and stocky, has been associated with the U.S. Air Force in a civilian capacity since 1951 and, before that, was active in commercial aviation.

Lowen and Bell quietly played sleuth and researcher. After hundreds of queries, a mountain of correspondence, searches of official files, and trips to South America, they have amassed what they believe to be the hard truth of Redfern's last hours in the air. He came down, they're convinced, not in the Guianas or Brazil—where thirteen previous expeditions were made—but in a remote and untrod section of Venezuela's Gran Sabana. In evidence, they have gathered long-ignored and overlooked reports, aerial photographs, eyewitness accounts, and references made from early—and confidential—oil company maps of the 1930s, all of which they carefully cross-checked and documented. Here is what they have uncovered.

After his meeting with the Norwegian ship, *Christian Krogh,* Lowen and Bell discovered, Redfern crossed the coast of Venezuela east of Caracas. He turned in a southeasterly direction and about three o'clock in the afternoon he was seen by the passengers and crew of the river steamer *Delta,* near Macareo on the Orinoco river. Eyewitnesses saw that it was trailing smoke. A little over an hour later, 110 miles up the Orinoco, American mining engineer Lee R. Dennison was sitting in the square of the sleepy town of Ciudad Bolivar when the populace suddenly came to life.

"Aeroplano! Aeroplano!" filled the air and people rushed to the river wall and pointed. Droning low along the river, heading from Macareo, was a high-wing monoplane. It flew steadily toward the town until it was within a few hundred yards of the town square, then suddenly turned left and headed south over open savannah country. Dennison noticed a thin trail of black smoke from the machine and assumed the pilot was having mechanical trouble and would soon land. The plane, Dennison also noticed, was a closed cabin type with a green fuselage and yellow wings, but it did not come near enough for him to see the numbers on its wings. Airplanes were still rare in that part of Venezuela and were always cause for much excitement. The following Monday he learned that an American aviator had taken off from Brunswick, Georgia, for Rio de Janiero and had been

feared lost in the Caribbean or the jungles of Guiana or Brazil. Dennison knew the plane he saw must have been Redfern's for there were no other airplanes unaccounted for in that part of the country.

Four days after he saw the plane pass over Ciudad Bolivar, Dennison set out in his Ford truck to deliver mining equipment to Paviche, about 130 miles southeast on the Caroni River. He and his assistant stopped for lunch at a small settlement on the Tacoma River. The native woman who prepared their food was very nervous. Dennison learned from her that she was upset because of the "Devil Bird" that flew over her house several days earlier. As he ate, Dennison questioned her closely.

"What did the 'Devil Bird' look like, Senora?"

"It was a big green bird with yellow wings, Senor," the woman jabbered excitedly. "I am afraid it will come back and steal my babies!"

"Did it make much noise when it came over your house, Senora?"

"Oh yes! It made a terrible noise, Senor! I'm sure it will come back and kill us all! It was so big!"

Dennison realized this was the same plane he saw turn south at Ciudad Bolivar; it had flown over this village, too. He tried to allay the woman's fears as tears streamed down her brown, plump face.

"You have nothing to fear, Senora," he told her. "That was an airplane and there was a man inside it." But the woman was not comforted.

Three days later, at Guri on the Caroni, Dennison asked Sute Menzez, the town's mayor, whether he had seen or heard anything of a green-and-yellow monoplane a few days earlier. Mendez shook his head; no airplane had flown over Guri, but between four and five o'clock in the afternoon on which Dennison had seen the plane at Ciudad Bolivar, Mendez and his wife had heard a steady sound in the air, toward the south, upriver. "It was like the noise which the motor in my launch makes when it runs away," Mendez said.

The times agreed, Dennison knew, for he saw the plane about four o'clock over Ciudad Bolivar and it would have taken less than an hour for it to reach the Caroni River. As Dennison pushed farther south to Paviche, he made other inquiries about the plane, but no one he spoke with had seen or heard it.

Working on the problem from several angles, then comparing and combining their independent finds, Lowen and Bell were able to obtain a copy of a lengthy report on the search for Redfern. It had been coordinated through the former Secretary of the Legation of the United States in Caracas. The report told of the findings of American bush pilot Jimmie Angel, who discovered the falls that bear his name, and it also included mention of a Bulgarian cartographer named Trayanoff, who was surveying farther south on an island in the Caroni when he saw the monoplane fly low overhead, trailing smoke.

A pattern emerged. By connecting the points where the plane had been sighted and heard, it became clear that Redfern had oriented himself over Ciudad Bolivar and had struck out on a new course—directly for Rio.

Now the Redfern saga took on a totally new perspective. And one thing was certain: Redfern did not get far beyond the island where Trayanoff sighted him. The Stinson was losing power—and the still-uncharted terrain was rising before it.

Further research brought the following to light: Dennison subsequently flew several times with Angel in the course of conducting his mining operations at far-flung camps throughout the Gran Sabana. During one of them Angel flew low over the jungle floor north of Auyentepui—Devil Mountain—to show the engineer a tangled airplane wreckage in the treetops. Angel told Dennison he had found it in 1935 and claimed it must have crashed before there were any other airplanes in that part of the country. Unfortunately, in writing about it later, Dennison was not explicit as to its location.

But Jimmie Angel was. Lowen and Bell obtained a chart

and were prepared to document the official report of the search. It showed the plane's location, according to a statement made by Jimmie Angel, to be a certain number of miles north of a Caroni tributary that runs east and west, and a certain number of miles east of the Caroni. While this certainly narrowed the area, it was not close enough to pinpoint. Then they interviewed Angel's copilot, who had also seen the wreckage and remembered Lee Dennison. The copilot said Jimmie Angel first sighted it by the glint of the sun on a piece of the broken windshield glass; parts of the wreckage were clearly seen—what there was of them—and there was no doubt it was a plane. The description of the terrain and many nearby landmarks coincided with the plane's location on the early map.

Since 1970, Lowen has made three trips to scout and photograph the jungle area, and make contact with persons in the Gran Sabana who could organize a sweep on foot. His latest trip was in 1978. Bell made one trip, using helicopter overflights and infrared photography to cover the area. Both investigators agree that the possibility of finding the remains of the old Detroiter—and its pilot—are highly encouraging, despite the density of the jungle growth and the passage of so many years.

What are their plans now? "To interest a sponsor," said Bell. "And not curiosity seekers. We've had enough of them already. With adequate backing we could go down there, form a party of natives and thoroughly and systematically sweep through the site where Angel, his copilot, and Lee Dennison saw the wreck."

"And it's still there," Lowen said. "If we don't get to it first, the geologists and surveyors will probably stumble over it in a few more years; they're slowly pushing into that area as they continue to develop parts of Venezuela. Of course, if that happens, the souvenir hunters will strip the wreck of everything that can be carried away. We're reasonably certain that no human being—Indian or Venezuelan—has even seen the wreck. Our contact down there tells us that the

Indians shun the area for some reason. They're highly superstitious, and it may be that they remember old tales of a 'Devil Bird' that flew over their fathers' or grandfathers' camps."

And if expedition fourteen finds the *Port of Brunswick?* "We'll return the plane—as intact as possible—to the States," Bell explained, "to a suitable memorial where the memory and spirit of Paul Redfern, and of those who believed in what he was doing, will be enshrined."

"Long before the airlines were able to establish routes between North and South America," Lowen said, "Paul Redfern visualized a much-traveled air corridor between the two hemispheres. It's here today—and he pointed the way. That's worth remembering, and it's worth going after the solution to the mystery as well."

Following her husband's disappearance, as the expeditions all failed, Gertrude Redfern gradually resigned herself to accept that Paul had perished on the flight. In 1937, after ten years of futile searching, elevated hopes, and saddening disappointments, she petitioned the Circuit Court of Detroit to declare her husband presumably dead. Today, Gertrude Redfern retains much of the charm and grace of the young girl who once stood on Glynn Isle Beach and waved to her husband in the green-and-gold monoplane fast disappearing into the mists. In 1969, she retired as administrative assistant to the president of a Detroit corporation which she joined in 1935, and now lives a still-active life in Ohio. In looking back to those few short years when the world awaited news of her husband's fate, she reflects today:

> Many articles have been written about the Redfern flight, and most of them have been more fiction than fact. But there's no need to go into that now, for much water has gone over the dam since Paul's ill-fated flight and one cannot live in the past. My life remains interesting and challenging and with my many outside interests I am always wishing for more hours in the day.

All of Paul's sisters are living. Until their death, his father and mother never lost hope. Paul was quite resourceful and didn't want us to give up hope if he had a forced landing in the jungle. Dr. Redfern was fervent in following the progress of the searches for this son; together we pored for hours over chart after chart of the Amazon River Valley. I think I know every town and village by heart. In all, thirteen expeditions were made, all with high hopes—but without success.

Of course, a number of stories have come out of the jungles about Paul's disappearance. Of these, the general consensus was that his crash was caused by a tropical storm.

Frankly, I believe we may never know the truth of what happened.

10

Flight to World's End

Australia's greatest air mystery began on the chilly, blustery Saturday of March 21, 1931, when the Avro-Fokker airliner *Southern Cloud,* with six passengers and its two-man crew, vanished without a trace somewhere between Sydney and Melbourne. It was also Australia's first major airline disaster.

Less than three years earlier, the airline's famous founders, Charles Kingsford Smith and Charles Ulm, with Harry Warner and James Lyon, took off from San Francisco aboard the Fokker trimotor, *Southern Cross.* They intended to cross the expanse of the vast Pacific and reach Australia with only two refueling stops—Hawaii and Fiji. With the disaster of the Dole Derby still fresh in the minds of many, the pessimists shook their heads. Even if they should reach Hawaii, the plane could never get to the Fiji Islands—much less to Australia. But on June 9, 1928, the largest airplane ever to land "Down Under" rolled to a stop in Brisbane. Kingsford Smith's successful flight opened a new era for Australian aviation. When the *Southern Cross* next flew the Atlantic, the reputation of the sturdy aircraft allowed Kingsford Smith and Ulm to establish ANA—Australian National Airways—and provide daily air service between the eastern Australian states. Kingsford Smith provided the technical and flying know-how and Ulm the managerial ability. Both wanted their dream of operating an airline in Australia to come true.

With headquarters in Sydney, the young company bought

Australian Consolidated Press, Ltd.

The Avro-Fokker airliner *Southern Cloud* disappeared on March 21, 1931, during a routine flight to Melbourne.

five trimotor Avro Tens of the same type and model as their *Southern Cross*. They named them the *Southern Sun, Southern Sky, Southern Moon, Southern Star,* and *Southern Cloud*—each in honor of the now-famous Pacific flyer. All were English-built by the A. V. Roe Company, under license of Fokker Aircraft in Holland. The airliners were powered with three Armstrong-Siddeley Lynx engines of 225 horsepower each. The aircraft design was among the first commercial types ever built to accommodate the pilot and copilot side by side. Australian civil air regulations did not require a second pilot, but ANA decided to provide one as an extra safety feature.

The accommodations were plush by 1930 standards. Each passenger had a window seat with a cushioned wicker chair and seat belt. The chairs were arranged in two rows that faced forward. The floor was linoleum-covered and the aisle was carpeted. Along the length of the passenger area was a long, narrow window of shatterproof glass, each section of which could be moved back if the passenger wanted fresh air. In cold weather the cabin was warmed from the engines. By today's standards, however, the airliner was little more than a noisy container with wings. It had no air or ground radio, but then, there were no radio beacons to follow or radar to lean upon anyway. There were only crude and untimely weather forecasts to follow, with navigation limited to dead reckoning and compass courses, and an occasional glimpse at landmarks below to confirm the airliner's position. Few passengers today would reserve a seat on such a plane, especially when it was to cross the most mountainous—and dangerous—country in Australia. But to the airways then the *Southern Cloud* was what the Boeing 747 is today. The press called it huge, a "giant of an airplane." By comparison, the Avro Ten's wingspan was 124 feet shorter and it carried only six passengers as compared with 442 on the Boeing. The 52,400 pounds of fuel carried by a 747 is fifty-one times the gross weight of the *Southern Cloud* at takeoff. Yet, in her day, the *Southern Cloud* was a handsome lady to behold, with her dark blue fuselage and silver wings, the "Royal Mail" insignia and crest on the passenger door, and her name boldly emblazoned on each side.

ANA was a singularly ambitious project. Not only was it the first flying operation to offer scheduled multiengine airline service between capital cities, but it was the only airline operating in Australia without a penny of government subsidy. Six months after the first trial runs, daily flights connected Brisbane, Sydney, Melbourne, and Launceston in Tasmania. But although air travel was beginning to prove popular with the public, the airline struggled with financial problems as the worldwide depression inched its way into the Australian economy. ANA was meeting its expenses— just narrowly—and looking hopefully toward better times when passenger revenues would increase. On that threatening March morning, if the company could have gone without even one day's receipts, they might have cancelled the Sydney to Melbourne run, but the orders stood and the scheduled flight was on.

At that time there was no weather service for pilots in Australia. As strange as it may sound today, the pilots got their weather information from the chart in the daily newspapers. The forecast for that day showed an extensive rainy area for the next forty-eight hours, covering most of eastern Australia from South Australia to New South Wales. And more rain and thunderstorms were expected along the return route from Melbourne to Sydney, with overcast and poor visibility.

Melbourne had fared worse than Sydney that morning. Rain and near-gale winds had pounded the city all night and heavy seas crashed on the beaches of Port Phillip Bay. Telephones were dead and the streets were flooded.

March was autumn time in Australia, an unsettled season that brought sudden changes in the weather. Over Mascot Airfield, outside Sydney, a blustery wind played with low-scudding clouds. Savage gusts battered the hangar walls.

"Smithy" had a preference for former Royal and Australian Flying Corps pilots. Many had been war aces. Pilots were plentiful, but Kingsford Smith and Ulm wanted only the best to fly the first airline in intercapital service. Among them were Eric Stephens, P. G. Taylor, dour George "Scotty" Allan, Travis Shortridge, and daredevil playboy Jimmy Mollison, the most unconventional member of the ANA pilots. Mollison later became a famous record-breaker of the 1930s; Allan became a world-renowned multiengine pilot. Dashing and picturesque, they were part of a time when flying men were young Galahads of the air—men with a singular magic and glamor reserved for the very few.

William Travis Shortridge was ANA's senior captain, thirty-three, a striking, Sandhurst-trained former British officer. He lived in Sydney with his wife, her daughter Elizabeth, and their own young daughter Yvonne. He was the most experienced pilot on the 450-mile Sydney-Melbourne run. In the cockpit of a small plane, Shortridge was a brilliant instructor and aerobatic flier. He had been a machine gunner in World War I and, later in India, he learned to fly. In 1927, he came to Australia where he joined ANA.

Shortridge was slightly tired. He had a cold, and the edge of weariness still lingered from the previous day's flight. He never complained about his flying schedule, however, although he was not slated for the *Southern Cloud* run that morning. Eric Chaseling was scheduled to make the Melbourne flight, but he had been sent to Broken Hill with serum urgently needed for a sick child. Shortridge took his place. Accompanying Shortridge was copilot and trainee Charles Dunnell who already had four hundred hours along the route with the senior pilots.

"Breakfast in Sydney, lunch in Brisbane or Melbourne" —a slogan confidently touted by ANA. To people of the great, sprawling continent, where the centers of each state were a day's travel by rail, it was a new and exciting idea. Seven passengers were booked for the Melbourne flight that morning but one, Jack Musgrove, cancelled because of pressing business in Sydney. And at the last minute Mr. and Mrs. A. Fraser of Melbourne exchanged their tickets for the Monday flight.

The passengers arrived at Mascot Airfield in the company's Studebaker parlor-coach. They clustered in the lee of the embarkation office to await boarding time. Two of the six passengers were women. Elsie May Glasgow, a large, well-groomed woman of forty, was the cook and housekeeper for Dr. and Mrs. Upjohn of Melbourne. She wanted to finish her Sydney vacation in an exciting way—by flying home.

The other woman was tiny, attractive Claire Stokes, twenty-five. She was an artist, with slender hands and dark brown eyes and hair. She decided to be with her close friend who was having an operation that, it was hoped, would prevent blindness. On the spur of the moment she had rushed to the Sydney Tourist Agency and bought a seat for her first flight. That morning, leaving Sydney's downtown ANA office, she rode with her friend John Watson in the company's coach to Mascot. In a beige wraparound coat with a fur collar she was chic and neat. A string of lapis lazuli beads and matching earrings added to her delicate beauty. John sensed she was apprehensive about the forthcoming flight and tried to distract her by talking about a party they had recently attended.

This was not Charles Clyde Bronson Hood's first flight. The stage producer for Sydney's Capitol Theatre had married the beautiful actress Bertha Ricardo three months earlier. She was starring in *Sons of Guns,* a lighthearted musical comedy at the Theatre Royal in Melbourne, and Hood had flown there every weekend since the play opened a month earlier.

Julian Margules, thirty, was an electrician who had specialized in the new sound motion picture projectors coming into use in Australia. From his Melbourne headquarters he often received rush calls to Sydney and Brisbane to repair the sound synchronization of the machines. ANA was the answer to his travel demands. In Melbourne, his wife and small baby awaited his return.

Although forty-two-year-old Bert Farrell was cautious about flying, he too realized the air service was an asset in his creamery company's expanding business. The Melbourne businessman had also bought a ticket for his brother Theo, who was not interested in flying at all. Bert offered the extra ticket to his cousin Fred, but Fred could not leave until Tuesday. Bert, however, could not delay; he had an important business meeting scheduled for Monday.

Unlike Bert Farrell and Claire Stokes, handsome Bill O'Reilly had taken to flying with a passion. As a rising young accountant, he welcomed the frequent calls to Melbourne, and the regular flights allowed him to plan for more time at his office. He tried to persuade his friend William Sheahan, a promising barrister, to fly to Melbourne with him. But Sheahan declined and, by so doing, lived to become the minister for several departments in the New South Wales government.

Its servicing almost completed, the *Southern Cloud* waited, a silver-and-blue ghost in the mists. Some who saw it head-on for the first time likened it to a giant moth. Several thermos bottles of coffee and a box of sandwiches for the passengers' in-flight refreshments were carried aboard. There was no steward or hostess, so the passengers helped themselves in flight. Sometimes, if the air was smooth, the pilot or copilot went back to serve them. Each passenger was given an envelope with two cotton-wool wads for ear plugs, and a few sweet mints to relieve airsickness. Thirty pounds of baggage per passenger was the limit, plus the mail bag. It weighed very little; air mail was new to the public.

Boarding began. While the passengers stepped one by one into the cabin compartment and found seats, Eric Stephens took off for Brisbane in the *Southern Sky*. As soon as the *Cloud*'s passengers were seated and strapped in the two rows of wicker chairs anchored along each side of the plane's interior, boarding attendant Bill Hamilton jumped down and shut the door. As the three engines were started and warmed, John Watson stood outside the plane at the opened window where Claire was seated. She waved and smiled nervously as the wheel chocks were pulled away and the plane inched forward. He stood outside the office and watched the plane lift off and disappear into the mist. It was eight-ten, he noted. Then, strangely, a heavy sadness came over him as he stared into the grey sky and heard the *Southern Cloud* fade from hearing. He told friends later, "I knew I'd never see Claire again."

The air route between Sydney and Melbourne was noted for its wild terrain and fickle weather that could turn ugly within minutes. A third of the course passed over the Australian Alps—the Snowy Mountains—that were often covered with clouds and squalls. Jimmy Mollison regarded the run as one of the most hazardous flying routes in the world because of the "dreaded Kosciusko range of mountains." In his book, *Death Cometh Soon or Late,* he wrote, "If the weather was at all bad the run was undoubtedly dangerous and it was seldom I completed the trip without having felt really frightened. . . . When we approached the impenetrable banks of clouds that surrounded Kosciusko I often wondered what my passengers would have said had they known that the pilot who hoped to see them through was only teaching himself blind flying by experience . . ."

ANA had two emergency airfields, with fuel supplies, for their planes on the north-south flights to Sydney and Melbourne. One was at Bowser near Wangaratta on the main Melbourne-Sydney highway; the other was a short distance farther south, at Benalla. Reg Stewart, in charge of the fuel there, wondered as he noted the strong headwinds the southbound plane must be bucking whether it could reach either refueling field.

Meanwhile, in Melbourne, Mollison left Essendon Airfield in the *Southern Star* for Launceston, Tasmania, a three-hundred-mile flight. Shortly afterward, Scotty Allan, with only three passengers bound for Sydney, took off in the *Southern Moon*. He climbed to nine thousand feet, above the foul weather, into a wind that blew in hundred-mile-an-hour gusts from the west-southwest. All the way to Sydney, flying on instruments and by dead reckoning, Allan was unable to find a single opening in the overcast.

The weather forecast in the newspaper was already a day old. It had said conditions would be "cloudy and unsettled generally, with rain and thunderstorms." Later each morning ANA usually telephoned the local weather bureau for an updated report, but if unusually severe weather was coming there was no way to warn the airliners already in flight. At ten-thirty A.M. Harold Camm at the Sydney weather bureau took the nine o'clock weather forecast from the wire. Conditions between Melbourne and Sydney were much worse than had been predicted earlier—the most severe weather in thirty years. Gale-force south-to-west winds, heavy rains, and thunderstorms with hail were waiting for the *Southern Cloud* when it pushed into the Australian Alps. At eleven A.M., Camm sent special warnings to all steamship lines and flying services. Then he telephoned Charles Ulm at the ANA office. Kingsford Smith was there too and, when Ulm hung up, they stared at one another, silent alarm in their eyes. The change in wind direction meant that the unsuspecting Shortridge, flying blind above the overcast, would have to fight sixty- to seventy-mile-per-hour headwinds and would be blown steadily eastward. The previous day's forecast of north-northwest winds which he was counting on had changed—drastically.

A short distance away in downtown Sydney, Mrs. Shortridge was at the pictures with her two daughters. Suddenly, in the middle of the feature, she was overwhelmed by an unreasoning sensation of near panic. Shaken, she took the girls home and telephoned Mascot. She was told there was no word from Essendon on the arrival of the *Cloud*. She called back again and again during the afternoon.

By noon, in Melbourne, relatives and friends of the passengers were anticipating their arrival. Bertha Ricardo went to the theater for her matinee performance, knowing that her husband would be waiting in her dressing room afterward. Gwen Margules was preparing a special dinner for Julian, and Mrs. Upjohn looked forward to May Glasgow's return. By three o'clock, Mrs. Upjohn was concerned about her housekeeper's delay and telephoned the ANA office at Essendon. They said the *Southern Cloud* had not arrived.

At about four-thirty, in a Sydney suburb, Charles Dunnell's father was puttering at odd jobs around the house. Suddenly he felt a strange sensation, deep inside, and had the clear thought of his son flying as copilot on an ANA plane. A bewildering thought kept racing through his mind: *Something's happened to Charlie . . . something's wrong . . .* He decided not to confide in his wife; Charlie's mother never wanted him to fly and it would only worry her.

At both terminals relatives were becoming insistent. What had happened? ANA officers tried to reassure them, but at five P.M., when the *Cloud's* fuel would have been exhausted, the company suspended their services and readied every available company plane for a search. The first inquiries showed that at several points along the early part of the flight route the *Southern Cloud* was seen by a number of ground observers—then it apparently disappeared with its human cargo.

Scotty Allan had landed the *Southern Moon* at Mascot about twelve-thirty A.M. Because of the brisk tail wind, he arrived from the usual five-hour flight well ahead of schedule. He reported his flight as "wild" but otherwise routine. No, he hadn't seen Shortridge in the air; there had been cloud cover all the way north.

At Benalla, Reg Stewart watched and waited all afternoon for the southbound airliner. He was certain the Avro-Fokker would need to refuel because of the strong head winds. But when dusk came he assumed the pilot had taken a different route and landed at another refueling stop.

About six o'clock, Charles Ulm telephoned Kingsford Smith at his home to tell him the *Southern Cloud* had not reached Melbourne. Smithy fought an uneasy feeling. "Possibly Shorty flew west, away from the weather, and landed out of reach of a telephone," he suggested cautiously. "Either that," Ulm theorized, "or he landed at Bowser to refuel and decided to wait it out." They were both wrong. And the news of the missing *Southern Cloud* was to spell *finis* for them and their young airline.

Early Sunday morning, Kingsford Smith was on the telephone at Mascot trying to piece together the flight path of the *Cloud*. It had been sighted over Mittagong, seventy miles southwest, at nine A.M. Then it was heard in the overcast above Goulburn, 120 miles away, shortly before ten o'clock. After Goulburn there were no further reliable reports, although a shepherd at Lambrigg station, south of Canberra, was sure he had heard a plane late in the morning. Pilots paid little attention; Shortridge was not in the habit of flying that far east, they said.

A check at Bowser and Benalla airfields showed that the *Cloud* had not landed there. They became the bases for the aerial searchers, the first of which were Captain F. W. Haig

SEARCH AREA FOR THE *SOUTHERN CLOUD*, VH — UMF
MARCH 23 – 31, 1931

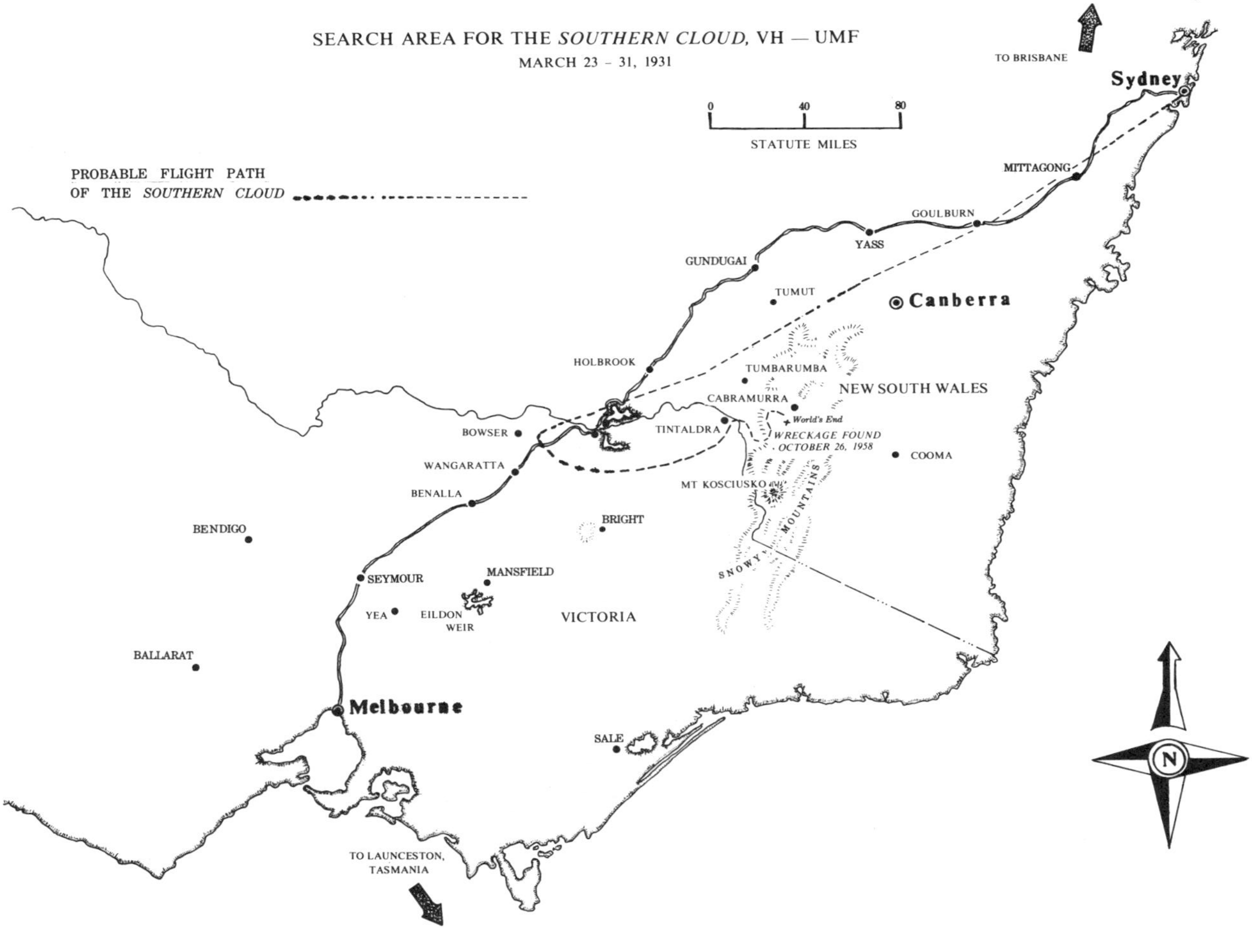

and Hughie Hughes in their Gypsy Moths. They swept east of Essendon, refueled, and searched toward the mountains before dusk closed in. From the Royal Australian Air Force station near Melbourne, two Wapiti biplanes joined the hunt. No organized search-and-rescue operation existed in Australia at that time; searches were made by friends or business associates and airplane owners. Although the RAAF joined in, there was no central or overall plan from a single directing source.

Mrs. Dunnell returned home from evening church about eight P.M. Suddenly she had the strong impression that she could see her son Charles at her side. She turned to her husband and blurted "Something's happened to Charlie!" A chill swept Mr. Dunnell; he had not told his wife he had had the same thought the previous afternoon.

Before the search ended on Sunday, Arthur Butler, in his Avro Avian, had teamed up with the searchers. Scotty Allan, after searching for three hours northeast of Melbourne in the *Southern Moon,* landed at Essendon to be greeted with a crowd of pallid-faced relatives. Before he could stop the engines, they pressed foward anxiously. "Did you find them?" they called. He shook his head.

After spending the night organizing a crew of observers for the *Southern Sun,* Kingsford Smith took off from Mascot at four-thirty A.M. in the face of a storm. By daybreak they approached Goulburn, then headed for Tumut, 270 miles from Sydney. Reports from the Victoria border indicated that several people must have heard or seen the *Southern Cloud* flying overhead during a violent thunderstorm on Saturday afternoon. Smithy crisscrossed the mountainous region, its turgid streams now swollen by rain. When he landed at Holbrook to refuel, the propeller of the center engine was damaged in a near nose-over. A new one was quickly flown to the searchers by Nancy Lyle, a young woman pilot.

By now the largest and most thorough air search in the history of Australian aviation was under way. Charles Ulm reached Benalla from Sydney in his Avro Avian and began to direct operations from there. Other pilots followed, one by one, flying their own or aero club planes. The consensus then was that the *Cloud* must have come down about one hundred miles northeast of Melbourne, in the Strathbogie Ranges.

So many reports now rushed in that it was difficult to unscramble the details. Oddly, however, none of them had been made to officials until *after* the plane was reported missing. Nor had one report included the plane's registration letters or the name *Southern Cloud* painted boldly on both sides of the fuselage. Some described it alternately as a monoplane and a biplane and such reports added confusion to the search plan. "Eyewitnesses" recalled seeing or hearing the plane "circle overhead" or "swoop low" or heard its engines "laboring." A woman and two children in an outlying settlement near Mount Separation claimed they saw the plane about four P.M. in a heavy thunderstorm. Nearby, a farmer claimed it flew low over his field about three-thirty in the afternoon. Then a minister and two police constables in northeastern Victoria—separated by thirty miles—each said they heard the plane at the same time: two P.M. Six people forty miles from Melbourne said they heard a crash and explosion shortly after five o'clock in a deep gully over which the ANA planes usually passed.

Although twelve planes were now searching from dawn until dusk, the call went out for more aircraft. Jimmie Mollison joined the fleet in the *Southern Star* as soon as he returned from Tasmania. The planes first concentrated on a mountainous and overgrown area between King Lake and Yea, which began forty miles from Melbourne. This seemed, from reports, to be the most promising area. The hollows between the deep, razor-backed peaks were overgrown with brush and littered with broken rocks. Often, they were filled with dense mists. Ulm and Kingsford Smith knew from their own earlier forced landing in the Kimberly district, during which search planes had passed directly overhead on three different days, that search pilots could fail to see a downed airplane. They had been down in open country; the *Cloud* was apparently down in a forest covered by a canopy of trees which made the search even more difficult.

With his load of binocular-equipped observers, Allan patrolled systematically in the *Moon,* flying north and south to cover a strip a mile wide, then crisscrossed it from east to west. He had cut a hole through the outer fabric under the cabin floor where one observer lay to scan the ground where the blind spots for those at the windows restricted the view. Food and medical packages tied to parachutes were stowed in the rear of the fuselage, ready to be pushed out when the *Cloud* was sighted. In the repaired *Sun,* Kingsford Smith flew east to Tumbarumba and south to Mount Kosciusko. Ulm, with others, continued to patrol from the muddy fields of Benalla and Bowser. Throughout the mountains ground parties were organized. From Wangaratta alone, hundreds set out by car, or by horse and on foot where the terrain was roughest.

At Essendon, distraught clusters of relatives continued to query the pilots when they landed. At the dismal and windswept airfield tensions ran high and mingled with a gnawing fear of the worst that could have happened to their loved ones—wherever they were.

On Tuesday, the widespread search gained new momentum as dozens of planes departed and returned. Then the weather turned foul. But at eleven o'clock Allan defied the elements and took off to search the hill country around Warburton and Gembrook. With Bob Boulton as his copilot, and eight observers, he pushed his big trimotor hundreds of feet down into the steep, dark ravines between towering mountain peaks shrouded in clouds. He weaved in and out of the heavily wooded ravines and skimmed the sides so narrowly his observers quailed at the nearness of the steep gorges and ravines. Except for a few hours in the middle of the day, deep shadows darkened the ground here, and it was impossible to see details along the immense clefts. Low in the Strathbogie Ranges, Kingsford Smith flew in the same desperate manner, and when he landed at Essendon, he was welcomed with the most encouraging report thus far. Frederick Fitzalen, a pilot, was fishing on Eildon Weir, eighty-three miles northeast of Melbourne on Saturday, and said he had seen the *Cloud* pass overhead at five o'clock. Kingsford Smith valued the observation of a trained pilot, and with Fitzalen and Bertha Ricardo, Clyde Hood's wife, he took off to scour the slopes of Mount Torbeck. But after two hours of close scanning of the suspected area in bitter-cold air, the search proved fruitless.

When the *Southern Sky* arrived in Sydney from its Brisbane run that day, mechanics quickly refueled it, and it too was loaded with emergency rations and medical supplies strapped to parachutes loaned by the RAAF. Hundreds of reports had now poured in through the telephone network. Squadron Leader A. H. Cobby, Australia's leading war ace, telephoned Kingsford Smith to say that, on Saturday afternoon, while he was in an east Melbourne suburb, he heard an ANA plane fly low overhead in the cloud cover. Others substantiated his report. Could Shortridge have overshot Melbourne in the overcast and crashed in Phillips Bay or the Bass Strait?

From the majority of persons interviewed, the tendency to inject nonexistent facts which became "the truth" was evident. Kingsford Smith remarked on the misleading and widespread reports on the *Southern Cloud*, "I'm astounded by the many messages that appeared to be genuine, but which had no truth whatsoever in them."

On Wednesday, five days after the *Cloud* was last seen, ANA officials knew that anyone who had survived the forced landing or crash would have to get help quickly. Possibilities now suggested that Shortridge had been blown eastward toward the Alps, where snow was already falling and the nights were bitter cold in the heights. All Australia was aroused to the hunt for the vanished airplane. Newspapers swelled with photos and accounts. Mountain settlements emptied as more search parties set out. A local Light Horse regiment was mustered. Unemployed workers formed search parties. And in the air twenty-five planes droned on, scanning nine thousand square miles. The single engine light planes, traveling in pairs for safety, combed the less hazardous country; the large Avro Tens continued to scan the mountains and ravines.

Kingsford Smith, Mollison, Ulm, and Allan forced themselves to the peak of exhaustion, flying nine hours each day, often landing after nightfall by the light of crude oil drum flares. On Wednesday evening, Kingsford Smith's party landed at seven P.M. in the darkness, after a 250-mile flight to snow-capped Mount Kosciusko. They had checked out a report from Tintaldra, forty miles from Kosciusko, that flashes were seen near there late Saturday. Mrs. Fred Thomas said she heard a plane—uncommon in that region—overhead at lunchtime on Saturday and that evening she and a neighbor saw flashes in the mountains for about fifteen minutes. Mrs. R. Roberson, the postmistress there, said she had also heard a plane around noon. Observers in the *Sun* swept the rock-scattered ravines of the Alpine country, over miles of ranges, dense and impenetrable with thick growth.

At dusk each day, despairing Elsie Farrell and her oldest child turned away from her vigil at Essendon. Each time the youngster held back, telling her mother they should stay and "wait for Daddy." Elsie Farrell drove frantically into the hills to search for clues and track down reports. She chartered a plane to make sweeps where new rumors had surfaced.

Day followed weary day, each more unreal than the previous one. The long flights through the dangerous mountain passes became nerve-wracking and nightmarish. When the pilots left their planes at night, sore and stiff-backed, their faces were pale and lined with weariness. They walked in a daze and collapsed onto their beds unable to sleep. Relatives besieged them, pleading for news; but there was no news. On Wednesday, Kingsford Smith, for the first time facing desperation, told reporters, "Yes, things are bad—but I'm not giving up hope . . . yet."

It was difficult for many to understand how a plane the size of the *Southern Cloud* could disappear without a trace. But those who were aware of the enormous spread of the country into which it vanished had no problem imagining what must have happened. The New South Wales-Victoria border country was virtually unexplored at that time. In its vast tracts were hundreds of sheer cliffs, thick forests, ravines, peaks, and gorges—much of which was and still is inaccessible. Even today only the most venturesome and hardy foot travelers willingly challenge the country. Railroads and highways remain sparse and much of the land is still viewed best from the air. In 1931, the search pilots saw only miles and miles of the rough and unbroken terrain, forbidding and impenetrable, where not just one, but hundreds of airplanes could be lost forever.

Near Tumbarumba in the Alps foothills, seven experienced bushmen searched one area of forty thousand acres that could be covered by three planes in an hour. Of more than five hundred reports, 383 were investigated.

On Thursday, an RAAF amphibian investigated reports of aircraft wreckage in Phillips Bay. It proved false. Number 3 Squadron planes made runs over Tintaldra, from which one of the more believable reports had come. The next day Kingsford Smith, drawn again to the Snowy Mountains, made a seven-hour flight with eight observers around Mount Kosciusko and the Alps foothills.

Some friends and relatives of the missing passengers organized small searches, but the Farrells got a large hunt underway. Mansfield in northeastern Victoria was their base, and they offered five hundred pounds for news that would lead to the plane's recovery. On Sunday morning, eight days after Shortridge left Mascot, the official search was called off, but Kingsford Smith told reporters that two ANA planes would keep going, along with three other smaller planes. The most widespread and thorough air search ever made in Australia had failed to uncover one clue as to the location of the plane and the eight people aboard. Up to that time, ANA had carried eight thousand paying passengers 671,000 air miles without the slightest accident, much of it over some of the most mountainous terrain in Australia, and with navigational aids that were sketchy in the extreme, compared with those carried on aircraft today.

ANA service resumed between Sydney and Melbourne on April sixth, and Scotty Allan slipped away to Braidwood, southeast of Canberra, for another look. One of the volunteer searchers, David Staig, continued to look for the *Cloud* for three years. In a high-wing monoplane he had designed and built himself, he flew hundreds of hours on a one-man mission during weekends and holidays.

Two relatives of the passengers, May Glasgow's sister and Bert Farrell's cousin, at separate times consulted a spiritualist to learn what had happened to the plane. Each was told the plane would be found in the wild Bogong district near the Victoria-New South Wales border.

On May eighth, a boy playing at Seven Mile Beach, south of Sydney at Port Kembla, picked up a piece of plywood on which was scrawled, "Whoever finds this piece of fuselage torn from the *Southern Cloud*. We are hopelessly lost. Compass done, Shorty." On the other side was another message, "God be with us and guide us safely. Shorty. Cheerio." On the edge of every disaster involving a lost airplane there seems to be a cruel hoax or two. This was one of them. Another occurred late in 1931 on the shores of Kingston Beach, Lacepede Bay, South Australia. A half-caste aborigine picked up a brown beer bottle containing a message. It said, "This bottle was thrown down from the *Southern Cloud*. We were lost and flying about, not knowing where we were. We were over water when we dropped this bottle." Tormented relatives, however, were unable to recognize the handwriting.

At first, in spite of the worldwide depression, ANA continued to build the valuable asset of prestige. It was a young, promising company that might have become a multimillion-dollar enterprise and put Australia on the world airlines map long before Qantas. The loss of the *Southern Cloud* and the heavy burdens and expense of the search changed all of that. The full force of the depression was now being felt; flying was a luxury people relinquished, as they did most traveling of any kind. It was evident by June that ANA was going to go under. The government could not help; its finances were thin and it would only assist those airlines serving the settlers isolated for months on end in the far outback. So, after only eighteen months of first-class operation, ANA's planes were withdrawn, one by one, from service.

The loss of the *Southern Cloud* was also the turning point in the lives of Kingsford Smith and Ulm. When the company went into liquidation they picked up what pieces they could and tried to continue with an overseas mail service, but they lost the *Southern Sun* when it crashed on takeoff at Alor Star in Malaya. The remaining Avro-Fokker Tens were sold to other airlines and the two fliers decided to return to international distance flights.

For ANA and its founders, bad news had become the order of the day. On the day the official inquiry into the loss of the *Southern Cloud* opened, Major C. W. A. Scott, a Qantas pilot, broke Kingsford Smith's England-to-Australia record in a 120-horsepower Gypsy Moth. Two days later, the pilot and passenger of the Avro Avian *Southern Cross Junior* were killed when its wing failed over Mascot.

Kingsford Smith remained, at heart, a flying pioneer when the pioneering days were ending and a new era for aviation was dawning. Within a decade, commercial aviation would be like any other business, with no more romance to it than operating a factory assembly line. Depressed by the loss of the *Southern Cloud* and his failure to find it, Smithy attempted another Australia-to-England dash. On September 24, 1931, he left Wyndham, Western Australia, in an Avro Avian named the *Southern Cross Minor*. Plagued with misfortune all the way, he fought weakness and illness until, half-dead, he was forced down in a remote part of Turkey. The suspicious Turks held him until all hope of attaining a new record had passed.

Back in Australia again, Kingsford Smith was knighted in June of 1932. For a time he took joy-riding passengers on local flights in his famous *Southern Cross*. In October 1933, he broke the England-to-Australia record in a Percival Gull named *Miss Southern Cross,* with a flight of seven days and four hours. Then, in December of 1934, Charles Ulm, George Littlejohn, and J. L. Skilling were lost at sea near Hawaii during their record-breaking attempt to fly from California to Australia in a twin-engine Airspeed Envoy, a fast and advanced aircraft for its day.

When Kingsford Smith made plans to form an Australia-New Zealand air-mail service in 1935, he changed his plans and decided instead to enter a race from England and Melbourne for the fifteen-thousand-pound prize. It was the Centenary Air Race sponsored by Sir MacPherson Robertson, a Melbourne chocolate manufacturer. Smithy's plane was a Lockheed Alstair named the *Lady Southern Cross*. On the night of November eighth, he and his copilot Tommy Pethybridge vanished over the Bay of Bengal, a stretch of water he had always disliked.

Oddly, every airplane that Kingsford Smith used in his long-distance flights—except the first one, the *Southern Cross*—was involved in a tragic end. The *Southern Cross Junior* crashed at Mascot and killed the pilot and passenger. The *Southern Cross Minor* crashed in the Tanezrouft of Central Algeria and its pilot died of exposure. The *Miss Southern Cross* crashed at Mascot, killed the passenger and badly injured the pilot. The *Lady Southern Cross* carried Kingsford Smith and Tommy Pethybridge to a watery grave in the Bay of Bengal.

New companies appeared on the air routes that were carved through the Australian expanse by Kingsford Smith and Ulm, and when the Douglas DC-2s and DC-3s appeared, regular service returned to eastern Australia. The Australian airline system spread across the island continent and, with Qantas in the lead, finally covered the world.

Meanwhile, the mystery of the *Southern Cloud* was a restless ghost; Australians did not quickly forget it. Years afterward, people continued to report "finds" and "clues" from the *Southern Cloud*. Letters continued to arrive at departmental agencies and rumors persisted. The plane was said to have carried diamonds and gold bullion; one of the passengers had thousands in banknotes in his briefcase. A camper found a battered vacuum flask in the bush; another found a handkerchief initialed "A. E. D." Others recalled seeing flares and fireballs in the mountains. On Mount Disappointment, a 2,631-foot peak in the Great Dividing Range north of Melbourne, some airplane wreckages were located. In 1954, the remains of an RAAF Wirraway trainer was found where it had crashed a year earlier. In 1955, in the same area, the remains of two U.S. Vultee Vengeance dive bombers were found where they had crashed and burned in 1944. But the mystery of the *Southern Cloud* remained.

In 1958, the Snowy Mountains Authority prepared to begin work on a four-hundred-million-pound dam project to divert the waters of two major mountain rivers for hydroelectric energy. In the great upheaval of earth and forest, the desolate valley known as "World's End" would be changed.

Tom Sonter, tall and curly-haired, was a young carpenter working with a subcontractor on the project to divert the Tooma River. On Sunday, October twenty-sixth, he left his construction camp fifteen miles from Happy Jacks, the highest village on the continent, to pursue his hobbies: hiking and color photography. To the south, amid a cluster of lesser peaks, he could see Mount Kosciusko, 7,314 feet, the highest elevation in Australia.

Sonter trekked for several miles, with no particular distance or direction in mind, into what may well have been an area as yet untrod by the first human. As he crossed a shadowy gorge, he stumbled suddenly upon a tangle of metal tubing, the same thickness as saplings interwoven among it. The pieces were twisted at odd angles, bent and shapeless. Careful of snakes underfoot, Sonter kicked at the fragments with his boots to free them from the leaf mold in which they were half-buried. More metal pieces came to view: brass and aluminum. He picked up several of the smaller pieces, but because the shadowy gloom cast by the overhanging trees made it difficult to search further, he soon left. Back at East Camp on the Tooma road, he told his fellow workers of his strange find and showed them the items he had picked up—a piece of metal with a serial number, a gas tank cap, and a stamped machine part. The men talked and speculated. Someone asked, "Wasn't there supposed to have been a plane lost here in the early 1930s?" Sonter hitchhiked to Cooma and caught a plane to Sydney, where Civil Aviation authorities asked him to lead them to the site. He did, to the steep slopes of the valley running east and west, a place of lonely silence, a quiet, tree-shaded slope. The officers scraped away the rotting leaves and uncovered a brass plate. It read, "Avro Type X DWG No. P11131." At last the mystery of the *Southern Cloud* had been solved—by a man not yet born when the plane took off on its last journey and who, on a million-to-one chance, meandered on this particular course through the mountains. But most astonishing of all, the *Cloud* lay on the southwest side of the densely-timbered ridge in the Toolong Mountains of the Great Dividing Range, only 250 feet from the top of a forty-five-hundred-foot ridge overlooking the Tooma River gorge and about twelve miles— in a straight line—from Kianda. It was in the Bogongs—as predicted by the spiritualist.

    The discovery of the plane brought new speculation as to what had happened to bring Shortridge, his copilot, and six unsuspecting passengers to this end twenty-seven years earlier. A close investigation of the surface provided most—but not every—answer. Beneath the soft earth the

Mirror Newspapers Ltd.

Recovery workers sifting through leaf mould and wreckage in the cockpit area of the airliner.

men uncovered the passenger cabin and cockpit. Piece by piece, they picked out the remaining relics: three watches, an aircraft clock, a tachometer, and an altimeter. One of the watches had stopped at one-fifteen. There were a string of chunky red beads from the thirties, a dented and rusted vacuum bottle that still held liquid, silver and copper coins, a key ring on which the name Clyde Hood could still be seen, a perfume bottle, a leather belt, dinner-jacket studs, a razor, and binoculars. In a pair of water-logged black shoes they found a few crumbling foot bones.

Sifted from the debris directly behind the center engine in the cockpit area were more bone splinters; everything else was ashes. All three engines had impacted deeply into the hillside, although the center engine was partly visible.

After inspectors had examined the entire wreck, they were able to put together the *Southern Cloud*'s last seconds of flight. The airliner plunged into the hillside, ripping through the dense underbrush as it struck. The single, tearing sound broke the silence in the valley of World's End as the plane exploded and burned. In less than an hour, with a final shudder of cooling metal, the wreckage settled itself into the charred ground and was still. Years slipped by. Weeds, brush and vines slowly covered the scar in the wounded ground, and trees took root amidst the tangled metal. The forest slowly closed over it and, in the silence of the Snowy Mountains, there were no clues to the buried secret in the valley. Almost three decades would pass before Australians would know what lay hidden at World's End.

The *Cloud* was lying on its right side with the front of the plane badly telescoped. Investigators knew it had caught fire on impact; the fuel tank metal had melted, the porcelain of a wristwatch had melted, and the few remaining bones of the occupants were calcified. The plane had been under control when it struck, during a turn, and was flying at near cruising speed, pushed by a tailwind of considerable force. Officials speculated that Shortridge had seen the hillside looming before him and, too late, had tried to veer away.

Now some of the reports of twenty-eight years earlier began to make sense. Many had apparently been valid but had largely been dismissed in the vast number of other reports. On that Saturday, a boy, his father, and another man were camped a few miles from Kiandra when they heard the *Cloud* pass low overhead about midday. The overcast was so low they could see nothing, however.

The day of the crash, Matt Bradley and Will and Harry Jenkins were prospecting for gold in the recesses of the Alps. A blizzard howled around them and, as the snow flurries cleared for a few seconds, they all saw a large plane pass slowly and unsteadily overhead, just above treetop level. It disappeared toward Black Jack, a 5,260-foot peak. They were only a mile from the crash site. A short distance from them, Tom Taylor was trapping dingoes when he heard plane engines to the east, near Yaouk Station. At the station, Mr.

and Mrs. Arthur Cochrane had heard a plane circling low, but they could see nothing in the driving blizzard. Two days after the *Southern Cloud* was reported missing, Robert Byatt told the manager of Marangle Station, twenty-six miles from Tumbarumba, that two shepherds, Dyer and Ross, who were camped three miles apart, had heard an explosion about one P.M. on the day the plane vanished. When the shepherds met later and compared notes, they had the same story: plane engines suddenly stopped, a brief silence, then an explosion. At that time no one in the district knew an airliner was missing.

Ironically, during the air search, planes flew over the crash site many times, yet no one saw the telltale patch of burned-out timber where the airliner lay. Kingsford Smith assuredly flew directly over the spot two days after the *Southern Cloud* had vanished. But his later theory as to the whereabouts of the missing plane was proved accurate. Shortly before he flew to his own death in 1935 he said he believed the airliner had crashed in the mountains around Kosciusko.

When the investigators declared their work ended, the souvenir hunters descended in droves on the twisted wreckage. With hacksaws, wrenches, and cutting tools, they worked round the clock by lantern light to carry away sackful after sackful of mementos and relics.

At Tintaldra, Mrs. Fred Thomas acknowledged that the air searchers had little chance of sighting the wreckage in the dark gorges near there. Almost thirty years earlier she had reported hearing the *Cloud* overhead, circling in the mists. Mrs. Roberson, the postmistress there, also remembered hearing the plane and making a report. She said, "We always thought it would be found there."

In Sydney and Melbourne news of the lost airliner's discovery was met with mingled emotions by relatives of the long-dead passengers. In Rose Bay, Sydney, Mrs. Y. A. Hayter, the daughter of Travis Shortridge, said, "Naturally I'm not happy, but it is a sort of relief. At least we know they didn't suffer. For years I suffered, particularly at night when I was a child. I used to imagine all sorts of things, as small children usually do. I was haunted with the thought that Daddy escaped death but was suffering from a loss of memory and no one was able to help him." Mrs. R. H. Stokes, the mother of Claire Stokes, admitted that she, too, was also tormented with the thought that her daughter might be wandering in the bush after the crash.

There was no coroner's inquest because there were no bodies to identify. Less than a bucketful of charred bones was recovered, and these lay in a storeroom of the Cooma police station for over two years. In December of 1960, about seventy people attended the victims' burial in a common grave under the shadowy pines of Cooma's cemetery.

The *Southern Cloud,* when found, was one of the world's few airliners that had flown on a regular route and a scheduled run still reported missing. Soon afterward, the people of Cooma, the nearest large town to the site, made plans to recover what remained of the plane and build a memorial to the pioneers of Australian aviation. Volunteers came forward and the Cooma Lions Club, the Ex-Serviceman's Club (aided by airline companies, technicians, and the government), designed a permanent memorial in which relics of the airliner would be displayed. *Southern Cloud* Memorial was dedicated on October 13, 1962. The striking structure has a concrete wing, forty feet in length, tilting upward from a triangle.

Nearby, under sheltering pines, the passengers and crew of the *Southern Cloud* are at rest under a stone engraved with the remark made by Sir Charles Kingsford Smith when, reluctantly, he abandoned the air search. It says, "Rest in Peace, Shorty—whatever the circumstances of the accident, I believe they were beyond human control."

11
"I'm Going On!"

The Sahara keeps its secrets well; still, it must relax its grip sometime. Proof of this was shown in French Algeria, whose scorched and waterless regions are totally unlivable. What happens to those who gamble single-handedly with the desert's vastness—and lose—was dramatically brought to light in the discovery of the body and plane of a Royal Air Force Reserve pilot in the untrodden Algerian wastes. On the night of April 13, 1933, ten years to the month before the *Lady Be Good* crashed in the Sahara, the pilot took off from Reggan in a small Avro biplane. His destination was Gao, a small town on the banks of the Niger River in what is today the Republic of Mali. Deep in the grim Tanezrouft, the loneliest and most terrible part of the Algerian Sahara, something happened, and the world lost contact with him for half a lifetime.

When the Algerian War flared in North Africa, French Army patrols ranged farther than ever before into the desert. One motorized patrol departed from isolated Reggan in February of 1962 and probed the sun-scorched uninhabitable region 150 miles to the south. Forty miles off the Trans-Saharan Road, in a stretch rarely seen by living beings, the motor carrier came to an abrupt halt. In their path lay strange skeletal remains. A closer inspection showed them to be the twisted, upturned wreckage of an old Avro Avian biplane.

Under the bare framework of a wing that had once offered some shade from the glaring sun, one of the soldiers found the huddled leathery corpse of a man. The mummified remains lay among empty water bottles and meager survival equipment: a flashlight, matches, some personal belongings.

Another man discovered one fuel tank that still held gasoline. Someone removed a weather-beaten, faded logbook that had been crudely wired to a wing strut. This was probably the pilot's final act before he died. It proved to be a diary that told the full story of the crash in painfully written paragraphs—eight days in the world's most merciless inferno. The brittle pages, remarkably still intact after weathering twenty-four years in the Sahara, revealed that the pilot had been Captain William Newton Lancaster, RAF Reserve. He crashed on April 13, 1933, while attempting a night crossing of the Sahara.

Lancaster was thirty-four at the time of the crash, a formidable, square-jawed aviator who had migrated to Australia as a young man and served with the Australian Light Horse in World War I. Later, he learned to fly with the Royal Air Force and soon won long-distance flight laurels. He was no fledgling when it came to spanning continents. During 1927 he and Mrs. Jessie "Chubbie" Miller teamed up for a successful five-month, thirteen-thousand-mile England-to-Australia flight in a tiny Avro Avian biplane, the *Red Rose*. Lancaster, though married and with two daughters in England, fell in love with the petite and attractive aviatrix. Mrs. Miller filed for a divorce, but Mrs. Lancaster refused to grant her husband his freedom.

Fame took Chubbie Miller and Bill Lancaster to the United States to promote aviation products, then came the crash of 1929. In 1931, they were living in Miami where Bill's frequent absences due to his work led to Chubbie's emotional involvement with Haden Clarke, whom she promised to marry. Lancaster was angered and, following an argument in the bungalow where they all lived, Clarke was found dead, allegedly by suicide. In the sensational trial that followed, Lancaster was narrowly acquitted of the murder of Clarke.

Lancaster returned to England where he realized the only way he could redeem his reputation was to make a record flight. There was now something about Lancaster that smacked of the desperate adventurer. Although unable to finance it personally, his parents scraped together enough money to buy an airplane. With it, the air adventurer made plans early in 1933 to beat aviatrix Amy Mollison's England-to-Capetown dash of four days, six hours, and fifty-four minutes. Lancaster's plane was another Avro Avian, formerly owned by Australia's most famous flier, Sir Charles Kingsford Smith, and used by him in 1931 for his unsuccessful speed dash from Australia to England. It was named the *Southern Cross Minor,* a light biplane with great load-carrying ability. With an added fuel tank it carried 115 gallons of gasoline and could remain airborne for fourteen hours. Its range in still air was sixteen hundred miles.

With years of long-range flying under his belt, Lancaster was not the least deceived by the magnitude of his venture. "I propose to fly over the inhospitable Sahara, a distance of fifteen hundred miles, in one flight. I've been warned to expect very bad weather on parts of the trip." It soon developed that his absolute determination to succeed had pushed aside rational thinking at a time when he needed it most.

At five-thirty A.M. on April eleventh, he kissed his mother and father good-bye at Lympne Airport in Kent and climbed into his heavily-laden blue plane. He took off on the first leg of his journey, circled heavily and slowly around the airfield in the dim dawn, and waved to his only well-wishers on the ground—his parents and a lone reporter from the London *Daily Express.*

Lancaster planned to fly nonstop to Oran, Algeria, on the first leg of his flight, but unfavorable winds over France forced him to land at Barcelona to refuel. The eleven-hundred-mile flight from London to Oran required sixteen hours. He landed in a state of exhaustion, rested as his plane was refueled, then prepared for immediate takeoff. French officials tried to detain him; they saw he was in no condition to go on. Stubbornly, Lancaster resisted. They reminded him that regulations required Sahara fliers to deposit eight hundred pounds to defray search costs. Lancaster did not have it. Irritated, distraught, he shouted, "I'll take my chances. I don't expect you to look for me!" The French colonial shrugged and stepped aside.

Lancaster flew directly across the stretch he feared most, the Atlas Mountains, in the pitch black of the predawn. At eight-thirty in the morning, he landed at Adrar, one hundred miles north of Reggan, where he decided to top off his tanks, overfly Reggan, and make directly for Gao. It was only eight hundred miles and well within the plane's reach. He was refueled and off in less than an hour, but as he headed south he encountered a sandstorm that reduced his visibility and blew him off course. When he picked up a road below, it led him to Aoulef, where he landed to get his bearings. Ten minutes later he was off again, headed toward the motor track that led to Reggan, his goal still Gao. But the strong desert blow pushed him back to Adrar, which he identified, and from which he turned again onto the course to Reggan. In three and a half hours he had flown only one hundred miles and he realized that he would have to land at Reggan to refuel before making the desert crossing. The sandstorm continued to rage.

Reggan was near the heart of the Sahara. In those days it was a post of the Trans-Saharan Company, an oasis that later became France's atomic test site. Lancaster had planned to make a daylight crossing of the desert, using the Trans-Saharan Road and a few landmarks to point the way to the Ahaggar Mountains in the south. If he arrived at Reggan early, he reasoned, he might be able to spend a few hours resting before a predawn takeoff for Gao.

The flying schedule went badly for the determined aviator. Head winds cut his ground speed to under seventy miles an hour. As a result, he did not reach Reggan until late in the afternoon of the twelfth. Then a threatening sandstorm clouded the skies around the airstrip. Lancaster made a quick check of his schedule while his plane was being

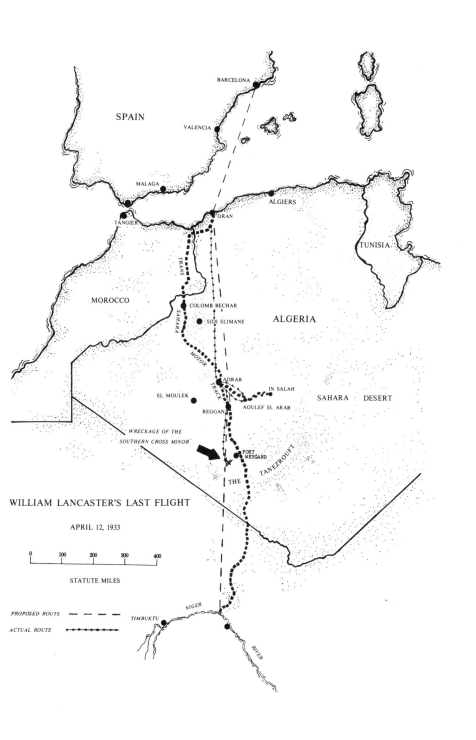

refueled. He was running well behind time. He knew at that moment there was only one way to break the record to Cape Town: to take off immediately on a night flight across the Sahara and navigate solely by compass.

The French officials at the post were disapproving when Lancaster told them of his decision. He had been in the air almost thirty hours without rest and could barely stand, much less fly and navigate simultaneously. They tried to dissuade him; reminded him that the weather was growing worse, that he had no cockpit lights, no reserve rations, and only one extra gallon of water—a day's supply in the desert. They also pointed out that with head winds his ground speed would be reduced to seventy or seventy-five miles an hour. He might not make it to Gao, even though the plane's range was sixteen hundred miles. The margin for error was too critical; it was clearly touch and go all the way. But with sleeplessness beginning to show its effects, Lancaster stubbornly refused to listen.

"I'm going on," was all he said.

The French official in charge of the post realized Lancaster meant what he said, so he shrugged his shoulders, sighed, and signed the clearance papers. Lancaster climbed clumsily into the plane's small cockpit and signaled for the propeller to be pulled through. The engine sputtered to life and, as the pilot was about to taxi away for takeoff, the Frenchman rushed up to the plane and handed Lancaster a box of matches and a flashlight. Lancaster nodded his thanks; now he would be able to see the compass in the darkened cockpit.

The takeoff was a mad grab for altitude; the lack of sleep was beginning to have its effects on Lancaster's reaction time. For several minutes the dazed pilot headed in the wrong direction, then, realizing his error, he turned south, on course. Slowly, the sound of the Cirrus engine faded into the distant darkness. The men of the barren desert outpost listened until they could hear it no longer. They who longed for civilization did not envy Lancaster's way of getting there. They turned back to their huts, arguing about the "crazy Englishman's" chances of reaching Gao. Some said he might make it if he could stay awake. In any event, radio confirmation from Gao in a few days would settle the matter. The message never came; they were the last people ever to see William Newton Lancaster alive.

The desert night air grew blustery and strong as Captain Lancaster headed south from the oasis. Cold blasts flew against his face. He welcomed the chill air; it kept him awake. He leveled off at one thousand feet and occasionally took a quick look at his compass with the flashlight.

Except for the rough air and the constant struggle to stay awake, the flight proceeded on course. Then, an hour and a half out of Reggan, the *Southern Cross Minor*'s engine began to act up. Immediately, Lancaster was wide awake . . . listening. There it was again—a misfire. He checked his power settings. Everything seemed normal. Sand perhaps . . . in the carburetor? A moment later and the engine cut out completely. In the lonely sky over the Sahara the only sound was the low whistling rush of wind past the struts of a small, powerless airplane and the muffled ticking of its windmilling propeller. Lancaster peered into the darkness below. He was determined to set the gasoline-laden biplane down in one piece. Then he could repair the damage and continue on.

But Lancaster's tired eyes played tricks on him. He misjudged the ground and the plane hit hard on the desert floor, bounced, hit again, and plowed into a mound of soft sand. The flimsy gear gave way, and the biplane flipped over on its back with a splintering crunch of wings and propeller. Lancaster pitched violently forward and slammed against the instrument panel.

Dazed and bleeding from face wounds, he crawled from the machine and surveyed the damage in the darkness. It was the end of the line for the *Southern Cross Minor*.

Half-stumbling, half-crawling, Lancaster burrowed under a wing where he slept until morning. He awakened with a compulsion to check his two-gallon water tank. He staggered to the broken fuselage. The tank was intact.

He took stock of his supplies. In all, things weren't too bad, he figured. He had two gallons of water, a thermos partly full, another thermos of coffee, and some food rations. Seven or eight days, he estimated. He looked at his map and judged he was about twenty miles off the Trans-Saharan Road.

He was wrong. Dead wrong. He was nowhere near the Trans-Saharan Road. Lancaster was lost, an insignificant speck in the world's largest unmapped desert.

He began to realize his weakness from fatigue and loss of blood. His head wound pained him and to pass the time he jotted down his thoughts in his logbook. He was confident his loss would be noted soon and searchers would start from Reggan that evening. He elected to remain with his plane; he had informed his parents and friends that this would be his plan if forced down. Too, he may have thought a passing caravan of friendly natives would rescue him.

He stripped fabric from the wings and made crude flares out of gasoline-soaked cloth wrapped around wire. He kept a lonely vigil all the next night, lighting the flares at regular intervals.

The nights as well as the days were agonizing. He wrote in his log:

> The contrast in temperature is ghastly. In the day it's so hot it's like being in an oven; at night I need every bit of clothing I have with me—vest, shirt, sweater (thick one), coat, flying jacket (light), muffler of wool, flying trousers over them, socks, underpants. In spite of all this, I am still chilly.
>
> The water tank gets cold at night, so by morning, if I fill the thermos flask, I have an ice-cold drink all day. I take a sip every half hour.
>
> Not getting lost in the desert much, I don't know the technical method for conserving one's water. I do know this. When I lift the flask for my sip I have to fight with all my will power to prevent my tipping it up and having a good drink . . .

As the hours passed, Lancaster became more concerned about the absence of search parties. As early as the night of the thirteenth, however, the French authorities at Reggan had sent a desert vehicle along the Trans-Saharan Road to look for flares. The following morning, two search planes left Gao, heading north. Other military aircraft based on the fringes of the barren wastes and at scattered outposts were alerted as the air-ground search was put into motion.

Lancaster's close relationship with Chubbie Miller had been renewed before the flight; she had helped him plan his route and flight schedule, she had made him promise to stay with the plane if he were forced down—and not try to walk to civilization. Now many of his thoughts were of—and to—her. In his log he wrote:

> Come to me Chubbie, but take care in the coming . . . Chubbie, remember, I kept my word. I "stuck to the ship" . . . and . . . Chubbie, give up flying!

Sometime during the fifth night, one of the ground convoys came maddeningly close to finding Lancaster. The dazed pilot wrote that he saw a flare in the distance. Elated, he struggled to his feet and lighted several torches, then sat back to await the party he was confident would soon appear. He even took the luxury of drinking a little extra water. But his celebration was premature. No one appeared. His lights were not seen.

The next morning he wrote:

> I am resigned to my fate. I can see I shall not be rescued unless a miracle happens . . . I have no food of course. Only feel thirsty.

On the sixth day, the effects of prolonged desert exposure and loss of blood from his wounds began to tell. His skin was drying fast; his face swelled and his eyes ached. Now his writing was becoming illegible and he concentrated on sitting in the sheltering shadows of the plane's wings. He tried to conserve his energy to hold out as long as possible. He wrote that the only living things he saw were a small bird and a circling vulture that decided not to stay. That evening he stripped almost all of the fabric from the *Southern Cross Minor*'s frame and made the final torches. These he set alight after dark. Doggedly, he still clung to a thread of hope. His writing now was of memories of earlier, happier times. He reminisced to pass the time and bolster his fast-fading spirits.

On the seventh day, he could barely move, but he managed to scratch out the final log entry. He bade his friends farewell, wrapped the book in a scrap of fabric, and tied it to a wing strut with a piece of wire. Then he lay down to sleep, knowing the next day's water was gone.

The morning of the twentieth dawned. Lancaster raised himself on one elbow and stared blankly at the far horizon through swollen eyelids. On a card, he painfully scratched some last words, early, before the sun rose in the heavens.

> So the beginning of the eighth day has dawned. It is still cool. I have no water. No wind. I am waiting patiently. Come soon, please. Fever racked me last night. Hope you get my full log.
>
> Bill

The desert claimed him sometime that day, or perhaps it was sometime during the night, when he slipped into welcome death—still waiting. The search party finally came, twenty-nine years too late, and found him in a position that appeared as though he were expecting them. No passing caravans, friendly or hostile, had passed this way since the day he crashed. The French soldiers who found him carried his body and his personal effects to the place of his last departure—Reggan—where he was buried.

In 1931, young Wylton Dickson was hoisted to his father's shoulders to watch Kingsford Smith take off from Adelaide for London in the *Southern Cross Minor*. In 1974, Dickson's friend, Len Deighton, was in Adrar, where he learned about the recovery of Lancaster's remains. Deighton was told the broken skeleton of the Avian was still baking in the desert. He relayed the information to Dickson, who organized a party of seven Australians and seven persons from the United Kingdom, with two Land Rovers, to recover the wreckage. When they started out in 1975, Dickson said, "My aim was to find the plane, have it rebuilt and sent to Australia, for this was the only Avro Avian biplane in existence that had been owned by Sir Charles Kingsford Smith, Australia's best-remembered aviator."

After three grueling days of zigzagging through the blistering Tanezrouft, they came upon the tangled mass. They dismantled it with painstaking care, labeled and numbered each part, and loaded the pieces on their trucks. They drove back to Reggan and there the Algerian police impounded it, but allowed the members of the party to leave the country.

After prolonged negotiations with the Algerian authorities, the wreckage was finally released in February of 1979. Plans had been made earlier to transport the pieces to adventurer Keith Schellenberg's home, a Scottish castle in Aberdeen, where it was to be restored to flying condition. Following that, it would be flown back to Brisbane, retracing the course followed by Kingsford Smith in 1931.

But an examination of the long-exposed wreckage told technicians the *Southern Cross Minor* would never fly again. Mr. E. P. Wixted, Librarian of the Queensland Museum, and the father of a member of the recovery expedition, explained:

> The aircraft will not be restored as it is the almost universal opinion of knowledgeable people who have viewed it that restoration would in fact destroy a great deal of the aircraft that is historic.

On February 11, 1980, the eighteenth anniversary of the plane's accidental discovery in the Tanezrouft by the

Foreign Legion, the monument to William Lancaster's ordeal in the desert was placed on display at the Queensland Museum. It remained in very much the condition in which the 1975 recovery expedition found it. It had become, in the words of E. P. Wixted: " . . . a symbol of the successes and failures of the pioneering of aviation, of disaster turned to triumph through courage."

12

Amelia Earhart

On the morning of March 19, 1937, a twin-engine Lockheed 10 began its takeoff run at Honolulu. Halfway down the three-thousand-foot runway, a wing dipped and the pilot frantically chopped an engine to straighten the veering plane. When the uneven shear crumpled the right landing gear, the heavy monoplane, carrying 1,150 gallons of gasoline, ground-looped in a cloud of dust and sparks. Near the hangars a small group of spectators waited tensely for the explosion. It did not come.

The pilot, a slender thirty-nine-year-old woman, cut the switches and jumped to the ground as her navigators, Harry Manning and Frederick Noonan, stepped shakily from the passenger compartment of the broken machine. Tiny Howland Island, 1,940 miles in the western Pacific, would have to wait. Amelia Earhart would not get there with her eighty-thousand-dollar "flying laboratory" in that condition.

In three months, America's "First Lady of the Air" was ready for a second attempt to circle the globe, this time along an easterly route to benefit from prevailing tail winds. With Noonan as navigator, she took off from Miami on June first, bound for California—the hard way. The mystery of their fate, destined to grip the imaginations of millions, has never been solved.

The flight down the east coast of South America was uneventful. For Noonan, an experienced Pan American Clipper navigator, it was a familiar run. After a brief rest at Natal, the pair jumped off for Africa. During the final hour of

Smithsonian Institution National Air & Space Museum

Amelia Earhart standing in front of the Lockheed 10 Electra.

the tiring nineteen-hundred-mile leg over the South Atlantic, Amelia Earhart disregarded Noonan's course correction and deliberately altered her course to the left, instead of right. Forty-five minutes later they crossed the continent's coastline at Saint-Louis, French West Africa. Because of the aviatrix's disregard of Noonan's compass correction, they were 163 miles off course for Dakar, their intended destination.

After putting forty-four hundred miles of African wasteland behind them, they pressed on to Assab, then to Karachi. Dodging and fighting the monsoons that threatened to ground them for months, they battled their way to Rangoon, Bangkok, and Singapore. It was between Bandung and Surabaya that Noonan found a malfunction in their long-range navigation instruments. Reluctantly, they turned back to Bandung, where Dutch technicians worked on the mechanisms for two days. When they reached Port Darwin, they unloaded their parachutes, crated them, and shipped them home. The next day they flew to Lae, New Guinea, the departure point for the most grueling leg of the flight, 2,556 miles to Howland Island.

Forty days of flying, twenty-two thousand miles, thirty stops, nineteen countries, and five continents began to exact their price. Pilot and navigator were on the verge of exhaustion. Added to this was a growing uneasiness about possible difficulty with their instruments on the long flight. There were no landmarks in the vast stretches of the South Pacific, and Howland was so small that dead reckoning was out. There was only celestial navigation and a special radio homing station to guide them in. The weather would have to be clear for Noonan to take fixes from the stars at night and the sun by day. After they were a few hundred miles out, they would be out of touch with Lae and too far from Howland for radio contact. But Amelia Earhart was confident that, if she could keep an exacting course all night, early the next morning they could home in on radio signals from the Coast Guard cutter *Itasca,* anchored off Howland for that very purpose.

Even before their takeoff from Miami, a fateful conspiracy appeared to threaten the pair. It amounted to a myriad of seemingly unimportant details that culminated in disaster and in the controversial mystery that followed. To begin with, Amelia Earhart had had the plane's 250-foot reel-type trailing antenna removed before the flight to avoid the trouble of reeling it in and out. At Lae, she and Noonan realized this was a mistake. The trailing antenna would have enabled her to contact the *Itasca* much farther out than would the small loop antenna. She wished she still had it; now that they were facing the broad Pacific, the fifty-watt transmitter seemed woefully inadequate. Then, too, no one had told them of another more powerful radio homing aid on the island. A Navy high-frequency direction finder had been installed to assist them, but neither Commander Thompson of the *Itasca* nor Richard Black of the Department of Interior had informed them of this.

Fred Noonan had trouble calibrating his chronometers. If

they read slow or fast, he could not get an accurate celestial fix. Fifteen seconds of error in the precision dials meant an error of one mile from their actual position. A one-minute error would steer them four miles off. His pilot would have to navigate much of the distance at night with the magnetic compasses, but even here a one-degree error in longitude would mean missing the island by forty miles. In the face of these uncertainties, Earhart and Noonan set out to find a sand and coral speck in the Pacific.

At five-twenty P.M., Friday, July second, eight hundred miles out of New Guinea and on course, Amelia Earhart radioed their position to Lae. Everything was going well. She began to feel better about the trip, despite her growing weariness. Noonan dozed intermittently. Both were confident the *Itasca* was waiting and listening, and that in the morning they would pick up its signals.

Aboard the cutter that night, everything hummed with expectation. Radio equipment on board and at Howland had been checked in preparation for the Electra to come within range. Coast Guard radiomen were prepared to broadcast weather and homing signals on the hour and half-hour by voice and code. As a final check of their transmitting power, they called San Francisco and, from a third of the way around the world, came the reply, "Receiving you O.K." The Electra, using its call letters, KHAQQ, was to report by radio at fifteen and forty-five minutes past the hour.

Shortly after midnight, the *Itasca* made its first attempt to contact the Electra, first by voice, then by key with their homing signal. At twelve-fifteen, there was no return call. The *Itasca* tried again at the twelve-thirty schedule. It also radioed the *Ontario* standing by near Samoa and asked if it had heard the Electra's reply. The *Ontario* said it had not. At twelve forty-five A.M., KHAQQ again failed to respond.

Transmissions continued from the *Itasca,* but by one-fifteen they had not raised the fliers. Commander Thompson was not dismayed; the plane was still an estimated one thousand miles out. His radio operators continued to broad-

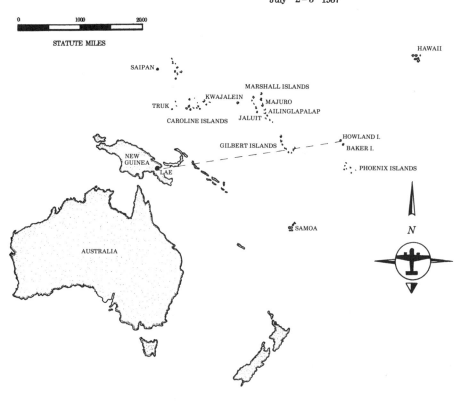

cast the weather on schedule, but their messages were not acknowledged at the Electra's scheduled reporting times.

Then, at two forty-five A.M., with the plane tentatively fixed at eight hundred miles, Amelia Earhart's voice was indistinctly heard through heavy static. The only intelligible words of her low monotone were "Cloudy and overcast..." —then, nothing. Spirits soared aboard the cutter. They had heard them! They were coming in! Frantically the excited radiomen worked to hold the contact. They checked their own signal strength again. Yes, the *Itasca* could be heard clearly by ships, including Japanese vessels, throughout the Pacific.

At three A.M., the *Itasca* transmitted the weather by key and voice. A pause, then they sent out the homing A signal "...dit-dah...dit-dah..."

No response from the Electra.

At three forty-five A.M., the aviatrix made voice contact, but the message was still garbled. She said she would listen for the homing signal. Again the *Itasca* tried to make contact but failed. It continued to transmit on the hour and half-hour but, except for another garbled message at four forty-five, the Electra was not heard again until six-fifteen, when Amelia called for a radio bearing. She said she would whistle into the microphone so the *Itasca* could get a steady sound. She added that she figured she was "two hundred miles out." Commander Thompson called the watch on the island and instructed them to get a bearing when the Electra transmitted. With two stations receiving—the *Itasca* and Howland—chances for a more accurate pinpointing of the plane should be good. The whistle went out, but the attempt to get a bearing on it was a failure. Then, at six forty-five, Amelia Earhart's voice, strong, clear, but with an unmistakable undertone of urgency, broke through the crackling static. "Please take a bearing on us," she begged, "and report in half-hour. We're about a hundred miles out!" Again the Howland direction finder swung for a bearing, and again it failed. This time she hadn't transmitted long enough.

At seven A.M., the *Itasca,* certain that the plane could not be far away, transmitted a fifteen-minute homing signal. Then, at seven-eighteen, the radioman called to inform the aviatrix that they could not take a good bearing on her. They asked if she wanted to take a bearing on them. They waited a full minute for her to reply. There was no acknowledgment. In the radio room of the cutter, the officers had the same question. Was Amelia Earhart having radio receiver trouble?

At seven-nineteen, the *Itasca* transmitted, "Go ahead on 3,105 [kilocycles]."

No reply.

Six minutes later they tried again.

No reply.

At seven forty-two, the woman's voice broke across the distance. "We must be on you but cannot see you . . . gas is running low. . . .Been unable to reach you by radio . . .flying at altitude of one thousand feet!"

The cutter acknowledged immediately. "Received your message, signal strength five [maximum]. Go ahead." Then it sent out the homing signal. It waited five minutes. No answer.

The *Itasca* repeated, "Your message okay. Please acknowledge." Nine minutes later Amelia Earhart called again, and by now it was obvious that she had heard none of the previous messages from the Coast Guard vessel, since she had failed to acknowledge any of them.

Her voice came through again. "KHAQQ to *Itasca.* We're circling, but cannot hear you. Go ahead on 7,500 now or on schedule time of half-hour." Her voice was loud and clear.

The Coast Guard radioman sent out a long homing signal on 7,500 kilocycles. Then, momentary success. She called back, ". . . receiving your signals, but unable to get a minimum. Please take a bearing on us and answer with voice on 3,105." She whistled into the microphone again while the *Itasca* and the Howland station worked frantically to get a radio fix. But the whistling could barely be distinguished from the mingled static and once again the attempt failed. If she had only counted slowly into the microphone, or if Noonan had transmitted with his key, the radiomen could have taken an accurate bearing. But the fliers failed to do so.

It was now eight A.M., Howland time. The fliers had been airborne for twenty hours. The men at Howland figured it was possible for the twin-engine plane to remain aloft for a maximum of twenty-six hours, or until two P.M. that afternoon *if* the pilot had leaned the fuel mixture sufficiently. Unfortunately, the waiting men had no way of knowing this.

At eight thirty-three, the *Itasca* called again and informed the Electra they were transmitting constantly on 7,500 kilocycles.

No response.

Considering the elapsed time from Lae, the plane should have been in the general area. The *Itasca* asked for a reply on 3,105. In a few moments, at eight forty-five A.M. to be exact, Amelia Earhart's voice broke through. It sounded broken, hesitant, yet urgent. Clearly, she was tired and very worried. "We are in a line of position 157–337. Will report on 6,210 kilocycles. Wait, listen on 6,210 kilocycles. We are running north and south!"

Nothing came through on 6,210. Suspense mounted in the *Itasca*'s radio room. The frustrated operators tried again and again to make contact, alternately transmitting and listening on all frequencies until ten o'clock. But the vast Pacific was ominously still.

Amelia Earhart had made her last call.

Radio officers and navigators tried to fathom the meaning of her last call, especially, "We are in a line of position 157–337." They assumed she had finally managed to take a bearing on them. "We are running north and south"—her final words—suggested that she did not know whether she was north or south of the line and was flying a zigzag search pattern for the island. That she asked the *Itasca* to take a bearing on her, appears to confirm their theory. Communications officers estimated from her signal strength that she was no farther than 250 miles, or closer than thirty miles, from Howland.

Commander Thompson acted promptly, realizing if the Electra were not already in the water, it soon would be. At ten-fifteen he ordered the *Itasca* north, reasoning that if the plane had flown to the south, it would have spotted Baker Island, thirty-eight miles away. The "337" line, he decided, was the one to search. If the fliers ditched without mishap, the Electra's empty tanks would keep them afloat, but they would not be able to use their radio; it depended on the right engine for power. Still, on the chance that they might yet be aloft, the *Itasca* continued to call, despite fading hopes that contact could be made.

In Washington, Admiral Leahy, Chief of Naval Operations, ordered four destroyers, a battleship, a minesweeper, a seaplane, and the carrier *Lexington*—with its full complement of planes—into action. During the next sixteen days, they set up search patterns around Howland and through the Gilbert and Marshall chains as well. They covered 250,000 square miles. Navy planes scanned 150,000 square miles and logged sixteen hundred hours. Planes and ships found nothing.

A stunned America listened at its radios, hoping that each news broadcast would tell of the rescue of "Lady Lindy" and her companion. In a few days the usual false reports began to come in. One of the search ships reported a green flare to the north. The missing fliers had flares, and when the *Itasca* veered north to investigate, it radioed, asking if Earhart and Noonan were signaling. If they were, it instructed, they should send up another flare. As though their message were heard, another green light appeared a few seconds later. Twenty-five witnesses saw it, and Howland reported flares in the same area. Another ship saw them, then they faded. The Navy decided that perhaps they had not been flares after all, but a meteor shower.

There were radio reports. All the way across the Pacific, from Hawaii to the West Coast of the United States, radio operators claimed they heard SOS signals from Amelia Earhart. Strangely, none of them were heard by naval radiomen on any of the Pacific islands or aboard the search ships, and so they were dismissed as imaginary.

Meanwhile, in California, a search of another kind was getting under way. It did not involve men, planes, or ships, but the unfathomable power of a gifted mind. Amelia Earhart's husband, George Palmer Putnam, called upon a close family friend, famed woman flier Jacqueline Cochran, who was later to head the WASPs during World War II. She and her husband, Floyd Odlum, had helped in financing Amelia's world flight.

The two women had often discussed extrasensory perception. During one of Amelia's visits, they heard that a plane had crashed somewhere in the mountains between Los Angeles and Salt Lake City. Miss Cochran gave a location, details of the terrain, and roads leading to the wreckage. Amelia Earhart checked with Paul Mantz, a friend and famed Hollywood stunt pilot. He confirmed the features on a map. She made a three-day search over the area, but failed to find the plane. That spring, when the snows melted, the wreckage was discovered two miles from the spot Miss Cochran had indicated. On another occasion Jacqueline Cochran found a missing airliner that had smashed on a mountain peak and was pointed downward. It was located exactly as she had described it. Again, she used her strange powers to record—and later had verified—the exact dates, times, and locations of stops made by Amelia and her husband during a trip in the Electra. It was inevitable that the two women should make a pact. If Amelia should go down and get lost, Jacqueline was to tell the searchers where to look.

It was a worried and distraught George Putnam who now asked Jacqueline Cochran, "Where are they?"

Calmly, she told him the fliers had been forced to ditch. She gave the location. Fred Noonan, she said, had injured his skull in the landing. The Electra was still afloat. Although she did not know the name of the ship waiting for them at Howland, she identified it by name as the *Itasca*. Then she named a Japanese fishing trawler in the same area. "Tell the ships and planes to go there," she said. "The plane is drifting, but I'll try to follow it." The Navy combed the area, zigzagged, and doubled back over it again. No luck. This proved to be one of the heaviest disappointments in Jacqueline Cochran's life. She never again tried to use her unusual gift.

On July nineteenth, the Navy called off the search, the cost of which now exceeded a million dollars. They had failed to find as much as an oil slick. Speculation and the following rumors, none of which appeared to be founded on fact, continued to flourish.

Amelia Earhart and Fred Noonan had inadvertently flown over Japanese islands that were being illegally fortified, and were shot down.

The fliers had been requested by the United States Navy to "look over" and photograph islands in the Japanese Mandate and had been captured as spies.

The two were in love and had used the flight as an excuse to escape from the world. They had landed on a remote Pacific island and were living happily together.

The second rumor found support from Dr. M. L. Brittain of the Georgia Institute of Technology, as well as others who were guests of the battleship *Colorado*. The vessel was one of those pressed into the search. As late as 1944, Dr. Brittain believed the aviatrix and Noonan were Japanese prisoners and would be freed by advancing U.S. Marines. "We got the definite feeling that Miss Earhart had some sort of understanding with government officials that the last part of her world trip would be over Japanese islands, probably the Marshalls," he said.

Someone else believed this rumor, too. She was Amelia Earhart's mother, Amy Otis Earhart. Although her belief had no official basis, she affirmed that her daughter was on a highly secret government mission and had been captured by the Japanese. In 1949, she said, "Amelia told me many things. But there were some things she didn't tell me. I am convinced she was on some sort of government mission, probably on verbal orders."

The deeper one probes into the Earhart mystery, the more the Marshall Islands keep popping into the picture. The accounts of two Naval officers are a case in point. Eugene Bogen, now a Washington, D.C. attorney, was Senior Military Government Officer there in 1944. He was in a position to hear much local history from the natives. Several of them told him that, in 1937, "two white fliers—one a woman—came down between Jaluit and Ailinglapalap atolls [southeastern Marshalls]. They were close by Majuro, where they were picked up by a Japanese fishing boat. Later they were taken away on a Japanese ship bound for Saipan.

"Elieu, a missionary-trained native, was my most trusted native assistant," Bogen said. "He said the Japanese were amazed that one of the fliers was a woman."

Former Lt. Commander Charles Toole of Bethesda, Maryland, confirmed Bogen's statement. He operated between islands in the Marshalls, and said, "Bogen's absolutely correct. I've come across the same story myself."

Was the Japanese ship bound for Saipan? Some significant discoveries occurred in July of 1944 on that little twelve-by-five-mile island in the Marianas Group. Thanks to the Japanese interest in cameras and photography, invading U.S. Marines found a photograph album filled with pictures of Amelia Earhart. It is a fact that she carried a camera on her last flight, but no album. If there were any connection between the last flight and the album, it means that either the film in the aviatrix's camera was developed after her capture, or the pictures were taken by the Japanese. When Dr. Brittain was asked about the Saipan album, he said, "Yes, I believe a definite relationship exists between the album and Amelia Earhart's disappearance."

Two Pacific combat veterans would agree with Dr. Brittain. During the Saipan invasion, Sergeant Ralph Kanna, now of Johnson City, New York, was with an Intelligence and reconnaissance platoon. His job was to capture and interrogate prisoners. One enemy soldier captured near "Tank Valley," carried a photo of Amelia Earhart standing close to a Japanese aircraft. It was sent through channels to S-2 Intelligence. Through the nisei interpreters, the prisoner explained that "this woman had been taken prisoner with a male companion." Eventually, "both had been executed."

Robert Kinley of Norfolk, Virginia, found another photograph of the aviatrix on Saipan. Probing ahead from Red Beach One, his Marine squad came across a house near a small cemetery. While Kinley, a demolition specialist, was inside checking for booby traps, he saw a picture on the wall. "It was of Miss Earhart and a Japanese officer," remarked Kinley. "It had been taken in an open field with hills in the background. The officer's fatigue cap had a star in the center." When the Marine was wounded soon afterward, the picture was lost with all his other belongings.

When World War II came to a close, there was another surge of rumors, claims, and denials and the U.S. Navy made an official statement. Amelia Earhart was not on a mission of espionage. She was not on a Naval cloak-and-dagger operation. Her Electra had not been shot out of the skies by Japanese gunfire. She had not been captured, held prisoner, or shot as a spy. The rumors still flourish, however, because of certain singular events that came to light after World War II.

They began when General Hap Arnold sent Jacqueline Cochran to the Japanese Imperial Air Force Headquarters after the surrender to report on the activities of Japanese women pilots during the war. She found many records on famous American pilots and uncovered several files on her late friend Amelia Earhart. Clearly, the Japanese government once had more than a casual interest in America's most famous aviatrix. Those documents cannot be located today. Not one department of either the United States government or the Japanese government claims to have them. They have vanished.

Next came the story that made the first breakthrough in plausibility. It began in 1960, when Operation Earhart was formed by three Air Force captains. They operated without government sanction or backing. Theirs was a private and independent organization formed to answer one question: What became of Amelia Earhart? Captain J. T. Gervais, a pilot with the 817th Troop Carrier Squadron, and Captain R. S. Dinger, Public Information Officer of Naha Air Base, Okinawa, had read Major Paul Briand's biography of the aviatrix, published in 1960. It contained a startling account that purported to explain her fate.

Major Briand's story came from a woman now living in California, Mrs. Josephine Blanco Akiyama, who was on the Japanese-held island of Saipan in the summer of 1937. A Chamorro native girl of eleven, she was bicycling along the harbor road at noon, delivering lunch to her Japanese brother-in-law, J. Y. Matsumoto, who worked inside the military compound. The area was under the strictest security, but she had a special pass to be admitted to the restricted zone. She had passed beyond the gate when she heard an airplane low overhead. She looked up and saw a silver twin-engine plane. It ditched in Tanopag Harbor.

When Josephine Blanco reached her brother-in-law, there was much excitement in the compound. They joined a crowd that witnessed Japanese soldiers bringing two American fliers ashore. One was a woman dressed like a man, in khaki trousers and a sport shirt. Her hair was cut short. The man was tall and thin, similarly dressed. Both were pale, weak, and exhausted. Their plane, if indeed it was the Electra, had come down in an illegal and extensive military installation, for Saipan, as well as other Japanese-mandated islands, was being heavily fortified.

In a few minutes a file of Japanese soldiers pushed the natives aside and led the couple away. At a nearby clearing in the jungle, a volley of shots cracked out. It was all over in a second. The squad returned without the prisoners; presumably the fliers were executed and buried on the spot. Mrs. Akiyama later stated, after having studied a picture of

Earhart and Noonan taken during the last flight, that they were the two she and her brother-in-law had seen on Saipan. At the time, she said, the Japanese would have killed anyone whom they suspected of spying on their secret operations. When the news services heard Mrs. Akiyama's story, they queried Japanese government officials. Was it true? Were Amelia Earhart and Fred Noonan executed on Saipan? The officials wouldn't say yes and they wouldn't say no. They did say, "Records pertaining to the Imperial Army and Navy were either destroyed by us or taken away by American officials at the end of the war. The report would be difficult to verify." Why, after twenty-three years, would the Japanese not want to admit the detainment and death of the Americans? Perhaps to admit to an incident involving Earhart and Noonan before the war would be to admit that, in direct violation of international law, they did indeed deliberately fortify the mandated islands. This would involve a shattering loss of face, especially since the Japanese had already testified at the Tokyo War Crimes trials of 1946 and 1947 that the air bases and installations were for cultural purposes and to aid fishermen in finding schools of fish. It seems unlikely that Josephine Akiyama would deliberately fabricate the story; she had no reason. Since she first related the story to Dr. Casimir Sheft, a Navy dentist, on Saipan in 1946, for whom she worked as a dental assistant, she has had many opportunities to capitalize on it. She did not.

Meanwhile, Operation Earhart was hard at work in other quarters, digging deeper to verify Mrs. Akiyama's story. Here's what they found.

Four Saipan natives remembered seeing a plane crashland in Tanopag Harbor in the summer of 1937.

They watched Japanese troops take its two pilots to jail after rescuing them from the bay. (This does not agree with Mrs. Akiyama's account but, instead, ties in with later discoveries.)

One native recalled being invited by the Japanese to see a white woman hanged. She declined.

The general consensus of the island natives is that the two

Americans were executed out of sight and hearing of the local population in order to keep the secret of Japanese activities secure.

While Captain Gervais was busy on Saipan, a team from Columbia Broadcasting System questioned Japanese Diet member Zenshiro Hoshina, who was chief of Japan's Naval Affairs Bureau at the time of the alleged execution. His reaction was one of surprise and shock. "I absolutely deny it!" he replied brusquely. "No such execution could have taken place without my knowledge and approval."

Some late information found by Major Briand did not agree with Hoshina's indignant denial. Briand heard from Gervais on Saipan; on the very day the former Imperial war lord voiced his denial—July 6, 1960—Briand stated in Los Angeles, "Captain Gervais has obtained Japanese photographs [the photo album pictures?] showing that Miss Earhart and Noonan were captured and killed on Saipan as spies. He also has affidavits from seventy-two eyewitnesses of this capture and execution." He continued from the Gervais letter, which gave the details of the fliers' imprisonment as well as the locations of their graves "untouched and undisturbed for twenty-three years, not even by war." The next day Captain Dinger, at a news conference at Fuchu Air Station, Japan, echoed Briand's and Gervais's charges.

Operation Earhart was becoming of interest to the Air Force as well as to the Japanese government. Gervais and Dinger were ordered to report to Fifth Air Force Headquarters at Fuchu to present their evidence. After evaluating the information, the Air Force stamped it "incomplete and inconclusive," and tucked it away in a classified file. Their action, coming as it did when Communist anti-American riots were being staged in Japan, was thought by some to have been designed to avoid incidents that could embarrass the Japanese at that time. Nevertheless, Operation Earhart was told to stop its investigation and the CBS team returned to the United States.

Much of Operation Earhart was talk, theorizing, conjecture, and countercharges. Although it took a step forward in clearing the way for later investigations, it did not establish that (1) the "Americans" were indeed Earhart and Noonan; (2) it was their Electra at the bottom of Tanopag Harbor; and (3) the bodies in the alleged graves were those of the missing world fliers. And still unanswered were the following questions.

What could have happened during the night of July second, to cause the fliers to head for Saipan instead of Howland?

How could Noonan, veteran of eighteen Pacific crossings, navigator of the *China Clipper,* have been so much in error?

Many persons involved in the final phase of the flight claim it could not have happened.

Curiously, this most famous woman pilot of all time was not an exacting flier. She flew mainly for the fun of it, or because flying was a personal challenge. Many of her record-breaking flights were happy-go-lucky adventures. There were times during her career when she failed to pay attention to instructions. She had a firm confidence, perhaps a touch of overconfidence, in her flying abilities. For example, when Noonan directed her to turn right a few degrees in order to hit Dakar, she turned left. As a result they were 163 miles off course when they reached the African coast.

Saipan lies 1,650 miles north of Lae; Howland lies 2,550 miles east. If the Electra's magnetic compasses functioned normally (and there was nothing to indicate they did not), Amelia Earhart would have had to have made an incredible course error of almost one hundred degrees. Even though the sky could have been completely overcast all during the night and prevented Noonan from making even one star fix, at dawn the sun would have come up off their right wing instead of over the nose—dead ahead. True, they were tired, but hardly so exhausted as to ignore the obviously misplaced position of the most reliable reference in the entire Pacific.

There are other arguments against the pair having flown

the Electra to Saipan. Almost eight hundred miles out of Lae, at five-twenty P.M., July second, Amelia Earhart reported their position by latitude and longitude. It was determined by a celestial fix. She said she was "proceeding on course," so if they altered course for Saipan after this time, when they were almost a third of the way to Howland, they would have to have flown west to reach the Japanese island. Then, too, when the Navy called off the search after having carefully swept the area northwest and southeast of Howland, they decided that the Electra's position of 157-337 could not have been a radio line after all, but a sun line. Considering the time and latitude differences between Saipan and Howland (factors in plotting a sun line), the numbers still come closer to the Howland position.

Chief Petty Officer J. D. Harrington is a Navy journalist and member of the Navy Institute. He does not believe the two fliers ever turned toward Saipan but, instead, that they got at least halfway to Howland, possibly as close as fifty miles. While researching information for his book on Japanese human torpedoes in World War II, Harrington investigated the log of the fleet tug *Ontario* at the National Archives. The *Ontario* was stationed halfway between Lae and Howland. It had been previously arranged that Miss Earhart was to signal the *Ontario* around midnight, when she was due to pass over the ship. At two-fifteen A.M., the commander of the *Ontario* ordered the ship to steam for nearby Samoa.

"The *Ontario*'s skipper must have been satisfied that they had passed over, otherwise he wouldn't have left his assigned position," remarked Harrington. "And if they passed over the *Ontario,* they wouldn't have had enough gas to make it to Saipan."

Harrington believes the Electra went down somewhere northwest of Howland. He speculates:

> They might have been saved if high brass in Washington hadn't interfered with the *Itasca*'s rescue attempt. Radio

contact showed that the plane was near the island. Commander Thompson set out to search in the most logical area they could have ditched in, but Washington called the ship back to the island to serve as a seaplane tender. Bad weather kept the planes from searching for several days. Washington then released the *Itasca* to investigate false reports of the plane's location [the green flares]. By the time it arrived back in the area where the plane probably went down, it was eight days later.

The *Itasca* heard nothing after eight forty-five, on any frequency, by voice or key. An hour earlier the aviatrix transmitted, " . . . gas is running low. . . ." She meant precisely what she said. If she had had another four hours' flying time—until noon Howland time—it seems unlikely she would have said this. Those who reject the Saipan story believe the Electra came down shortly after eight forty-five. They point out the two-hour time difference between Howland and Saipan. Eight forty-five Howland time is six forty-five Saipan time. This does not support the Saipan natives' story that the plane pancaked in Tanopag Harbor about noon.

"During the summer" covers three months; "1937" could have been 1936 or 1938. No doubt Saipan natives witnessed something, but what? During the Pacific struggle, many planes must have ditched in Tanopag Harbor. After twenty-three years the memory dims; events run together.

For those who waited in vain on Howland Island, however, the last flight of Earhart and Noonan remains a vivid memory. Dr. David J. Zaugg, a medical officer of the San Francisco Merchant Marine Hospital, recalled:

> She was due to reach Howland around seven A.M. It was a bright, sunny day and we were waiting on Howland to service her plane when it landed. We began to pick her up quite clearly on the radio, but the trouble was, she made voice transmissions when we needed Morse code to

triangulate her position. She was flying directly into the morning sun and might not have seen Howland at all. I think she just went into the drink.

I never knew of any espionage work she was asked to do, and I never put any credence in the statements of the Saipan natives. It seems highly improbable to me that they [the fliers] could have gotten to the Marianas.

Another mystery: Why did Noonan not use his telegraph key?

Two other retired Naval officers are in accord with Dr. Zaugg. Rear Admiral Hoeffel was gunnery officer aboard the *Lexington*. "Regarding her possible landing on another island—there simply aren't any. I've never accepted the idea that she may have been captured and executed by the Japanese. I've always believed they crashed and drowned." Vice Admiral Beary, then executive officer of the battleship *Colorado,* said, "It's been my guess that Amelia Earhart ran out of fuel, crashed in the Pacific, and drowned."

When asked about the secret mission, Paul Mantz, who outfitted the Lockheed for its world dash, replied:

> I won't say yes or no. It was so long ago—it makes a man think hard about it. I installed one large vent which was channeled to all the gas tanks. If the plane pancaked on the water in one piece, all they had to do was pull a lever to seal the tanks for flotation. It could have floated indefinitely. They had all the survival gear they needed for a mid-ocean rescue: a rubber life raft, two-way radio, and other items. At the last minute she left some of it behind—I don't know why.
>
> Naturally I've wondered about what happened to them. Perhaps they didn't have enough gas to find Howland. . . . Perhaps Amelia Earhart's still alive. . . . I don't know.

Late in 1961, there was a renewed flurry of interest in the mysterious mid-Pacific disappearance. From shallow unmarked graves on Saipan, Fred Goerner, a San Francisco radio newsman, sifted bone fragments and teeth of what he believed to be the remains of the two fliers. Goerner, a most persistent sleuth in the Earhart puzzle, has made three trips to Saipan for CBS. During his 1962 trip, he stopped off at the Marshalls and found Elieu, the native who had given Bogen and Toole information about the American fliers. Now a schoolteacher on Majoro, Elieu told Goerner the same story.

Goerner reasoned that there might be a few natives still on Saipan (during his first trip in 1960) who would have remembered Amelia Earhart if they had once seen her; workers at Tanopag Naval Base, natives who had worked for the Japanese military police at Garapan garrison. He was right; after questioning almost a thousand Saipanese with the help of Monsignor Oscar Calva, Father Arnold Bendowski, and Father Sylvan Conover of the island's Roman Catholic Mission, Goerner pieced together an intriguing story.

In 1937, two white fliers, a man with a bandaged head and a woman with close-cropped hair, dressed in man's clothing, were brought ashore from Tanopag Harbor in a motor launch. The man was taken to a prison stockade at Punta Muchot; the woman was taken into town (Garapan) and kept at a hotel where political prisoners were detained.

The woman was kept under close guard. A native woman testified that, as a young girl, she lived across from the hotel and saw the woman prisoner almost every day when she was allowed a short exercise period in the yard. About six months after she was brought to Garapan, the white woman died of dysentery. Goerner found a native laundress who had worked for the Japanese and who had "many times washed the clothes of the white lady." Another woman, who worked at the Japanese crematorium near Garapan's native cemetery, said she saw the body of the woman being taken to a place just outside the cemetery along with the man who had also been imprisoned. There the man was beheaded and both were buried in a shallow grave. Goerner's assiduous search even ferreted out the woman whose father had supplied the black burial cloth for the dead woman, and a former dentist whose Japanese-officer patients talked too much about the "two American-flier spies."

With the help of native divers, the Navy veteran went down for parts of a two-engine plane in thirty feet of water. He recovered a generator he assumed was from the Electra. It was slime- and coral-encrusted from many years in the water. In the States, Bendix Aviation officials and Paul Mantz examined it and identified it as a Japanese copy of an American generator.

After Goerner returned to the States, he continued to track down leads. He got a tip from Thomas E. Devine of Westhaven, Connecticut, who was stationed on Saipan after it was captured from the Japanese. The natives told Devine of a white couple held captive there in July of 1937. The woman allegedly died of dysentery and the man was beheaded. According to a native woman, both were buried at Garapan Cemetery. In fact, she showed Devine the unmarked grave of a man and a woman who had come before the war. This was in 1945. Testimony from nineteen other islanders supported the woman's story.

Goerner was certain his months of work were about to pay off. He believed, however, that the pair had actually crashed near Jaluit Island in the Marshalls, where the Japanese were constructing illegal fortifications. It was Commander Paul Bridwell, of the Navy Administration Unit on Saipan in 1961, who convinced Goerner. Bridwell admitted to the newsman that a Naval Intelligence officer who followed up Goerner's earlier leads was unable to find flaws in the natives' statements. And Bridwell had a theory of his own.

> I think they went down near Alingalapalap, Majuro, and Jaluit atolls in the Marshalls. A Japanese supply ship took them to Yap in the Carolines and probably a Japanese naval seaplane [the one Mrs. Akiyama saw ditch in Tanopag Harbor?] flew them on to Saipan—that's why some witnesses think they came from the air. Back in

1937, four U.S. logistic vessels were supplying the Far East Fleet. The *Gold Star, Blackhawk, Chaumont,* and *Henderson.* I understand they intercepted some coded Japanese messages.

In Bridwell's opinion, their radio logs should make fascinating reading. Goerner tried to get copies of them. The Navy informed him they were among records declared missing or destroyed.

Why would Earhart and Noonan be sent to Saipan, two thousand miles to the west? Possibly the Japanese garrison at Jaluit did not want the responsibility of disposing of the white "spies."

The six and a half pounds of bones and skeletal fragments were taken from a grave at Garapan Cemetery and sent to San Francisco. University of California anthropologist Theodore McCown examined the bone fragments and thirty-seven teeth. He undertook the task with no preconceived opinions. Until he reconstructed the skeletons from the bones, Dr. McCown discussed the case with no one who had known the aviatrix personally, nor did he care to see photographs or become in any way personally involved. McCown said that in some cases it was possible to identify individuals from only one or two bones, but he made no such promise in this instance.

A tense week followed as the world awaited Dr. McCown's decision. It came—in the negative. His careful analysis showed the remains to include teeth from four individuals and bones from two. The tooth structure, the doctor revealed, showed them to be from Orientals.

Interest in the mystery of Amelia Earhart has not diminished. On the contrary, the ever-provocative question of what happened to the round-the-world flier refuses to fade away.

When Georner's book, *The Search for Amelia Earhart,* was published in 1966, it contained the accounts of two former U.S. Marine Pfcs.—Everett Hensen, Jr. of Sacramento, California, and Billy Burks of Dallas, Texas. They told Goerner of exhuming two skeletons from an unmarked grave outside Garapan Cemetery in 1944, under the instructions of a Marine officer—later identified as Captain Tracy Griswold. Henson asked the officer, "What are we looking for, sir?" The officer replied, "Have you ever heard of Amelia Earhart?" The captain placed the remains carefully in containers and took them away. According to a report in the Napa, California, *Register*, which had been working with Goerner on the investigation since 1963, an unnamed person in the government told an investigating newsman that the remains were taken to "the Washington area" and suggested they might be in "a rather obvious place." Arlington National Cemetery records were carefully checked, without success. Goerner asserted they were either secretly interred or were in the possession of the Armed Forces Institute of Pathology. He quoted Admiral Chester A. Nimitz, who told him in March of 1965, "I want to tell you Earhart and her navigator did go down in the Marshalls and were picked up by the Japanese." Unfortunately, Nimitz told him no more than that; he died the following February after trying, unsuccessfully, to have Marine Corps General Harry Schmidt—who commanded the 4th Marine Division at Kwajalein and Saipan—give Goerner more details.

In July of 1966, a spokesman for the State Department said its files showed no evidence either that Amelia Earhart was on an Intelligence mission or that she was captured by the Japanese.

Amelia Earhart's unfinished flight spurred Mrs. Ann Pellegreno, a pilot and former schoolteacher from Saline, Michigan, to complete the interrupted journey in 1967. With a three-man crew—copilot, navigator, and mechanic—she retraced Amelia Earhart's 1937 flight in another Lockheed 10 that was modified for long-distance overwater flights. As they flew over Howland Island, she dropped a wreath on its deserted beach.

Then, for a short time in 1970, the publication of *Amelia*

Earhart Lives, written by former Air Force Colonel Joe Klass, stirred hopes that the mystery would at last be solved. Klass, following a ten-year investigation with [now] retired Major Joseph Gervais, claimed—quite erroneously—that the long-missing aviatrix survived eight years as a captive in the imperial palace in Tokyo and was living in New York under the name of Irene Bolam, a widow, formerly of Monroe, New Jersey. Mrs. Bolam had, in truth, been a pilot and a former acquaintance of Miss Earhart, but she denied publicly and firmly that she was the famous flier. Equally erroneous was Klass's claim that a twin-engine Lockheed that had crashed on a California mountainside in 1961 was the long-missing Electra. The plane was, in fact, a Lockheed model 12, not a model 10, and its airframe and engine serial number proved it could not have been Amelia Earhart's aircraft.

In 1977, however, the thread of an earlier investigation resurfaced with startling results. Armed with information gathered over a decade, a group of investigators in suburban Cleveland, Ohio, prepared to bring suit in federal court under the Freedom of Information Act. Their aim was to force the government to release all of its records on Earhart and Noonan's disappearance.

Don Kothera, spokesman for the group, said, "We want to see that Amelia gets her day in court." He told reporters how he, with others of their group, found new pieces to the puzzle after interviewing Saipan natives. One woman, Mrs. Anna McGoofma, had related her experience as a seven-year-old schoolgirl—an experience that gave her nightmares afterward. She was walking home from school one day when she saw two Japanese soldiers forcing a white woman and a white man to dig a grave near a cemetery. Quickly, she darted behind a tree, fearful that she too would be caught by the soldiers. As she watched, she saw the soldiers force the white man to bend over—then he was beheaded by one of them. She ran before she could see what happened to the woman.

Mrs. McGoofma recalled the precise spot and took the investigators there. When they excavated, they unearthed bone fragments and some dental bridgework. The men brought the pieces—eighty-nine in all—home in a camera case to be examined by Ohio State University archeology professor Dr. Raymond Baby. His findings: The grave had held the remains of a white male and the largely cremated remains of a female who, he surmised, was "probably white" and just beginning to age. The Japanese were known to cremate those who had been ill. The dental bridge dated from the 1930s. Professor Baby, who retained the bone fragments, said, "They [the investigators] have a beautiful circumstantial case.... I have urged them to pursue the matter."

They did. The assorted bone fragments that were left behind strongly suggested the grave had been opened earlier and, in all likelihood, was the same burial from which Marines Henson and Burks had taken the larger bones twenty-four years earlier. Like Goerner before them, Kothera and others in the group searched out the two men who took part in the strange and grisly wartime detail. Henson and Burks again confirmed that Marine Intelligence Captain Tracy Griswold had—with notes or a map to assist him—shown them where to dig. The ex-Marines also recalled several other details of their work; they had found a rib cage, part of an arm, and the two skulls of the grave's occupants. Griswold, they said, had cautioned them to say nothing about the unusual afternoon mission.

When the investigators found Griswold, he told them that, on the record, he didn't remember anything. Off the record, however, he said they were on the right track.

In 1979, another Earhart investigator—Orlando, Florida, businessman Vincent Loomis—revealed the findings of his search. Unlike most of the earlier investigators, Loomis did not believe Amelia Earhart was on a spy mission when she and Noonan vanished. Loomis said he was in the Marshalls, on duty with the Air Force in 1952, when he discovered an airplane covered with jungle growth on a tiny atoll. He thought little of his discovery until 1967, when he read a book about the disappearance of the two fliers. After ten years of research, during which he and his wife made several trips to the Marshalls, Loomis was convinced the mystery was close to solution. "The pieces are falling into place," Loomis told reporters in Honolulu. "We are getting very close to solving the Earhart mystery."

During one trip, Loomis interviewed Biliman Amran, a Majuro businessman and former medical corpsman with the Japanese Navy in July of 1937. Amran said he had been ordered to board a cargo boat to treat an injured white man. He also saw a white woman whose description, Loomis claimed, "fits Amelia from head to toe." The most significant of Loomis's statements, however, was that witnesses told him of seeing the woman pilot and her navigator bury a silver container on a Marshall Islands atoll shortly before they were taken into custody by the Japanese. "I haven't the slightest idea what it may have contained," Loomis said.

Here the mystery stands. The vast Pacific still holds fast to the key, and speculation still flourishes.

Amelia Earhart Light—a lonely memorial lighthouse—stands on Howland Island. The Japanese Navy shot the top off during the war and it was never repaired. Once a year a Coast Guard cutter visits Howland. In 1962, crewmen from the *Buttonwood* painted the base of the structure white, to serve as a day navigation marker. A plaque on its foundation reads: AMELIA EARHART, 1937.

Aside from Charles Lindbergh, Amelia Earhart was the most famous American flier of the twenties and thirties. She was the personification of feminine emancipation—without the loss of feminine charm. An admission by the U.S. government that it had allowed the national heroine to embark on so perilous a spy mission would, even now, shock millions of Americans. The Japanese, for their part, would prefer not to have the world know they secretly executed a woman much admired by all nations—including their own. As allies today, the two governments would naturally cooperate in covering past incidents that could reflect discredit upon either.

"Someday," Amelia once told her husband, "I'll probably get killed. There's so much fun in life and so much to do; I don't want to die. But when I do, I want to go in my plane—quickly."

Perhaps she did; perhaps she did not. Until there's more definite evidence—or an official admission—we shall never know.

PART THREE

Famous Controversies

13

The Great Floating Palace

Air transportation came to a grim crossroads on May 6, 1937. The path it took on that fateful day at Lakehurst, New Jersey, eventually broadened into the global airlanes paced by today's sleek six-hundred-mile-an-hour passenger jets. If the *Hindenburg* disaster had not occurred, world aviation would have continued to develop the great rigid airships—the dirigibles. But in the twilight of that spring evening, the death of the airship era was close at hand as disaster stalked the giant that droned softly over the Jersey pinewoods. The mammoth LZ-129, mightiest and most successful lighter-than-air transport ever built, was about to die.

As it drifted majestically toward the mooring tower, none of the thousand spectators knew they were about to see the end of an epoch. The *Hindenburg* would burn into the public's mind an indelible picture of what could happen to people who traveled on a hydrogen-filled airship.

Although there had been a number of dirigible disasters since 1921, the *Hindenburg*'s safety record was flawless. In fact, not one paying passenger had, as yet, died in a dirigible disaster.

The *Hindenburg* was the most luxurious airship ever built, the pride of Nazi Germany. Its imposing form measured more than eight hundred feet. Nothing on America's airways could match its performance or range, eighty-eight hundred miles nonstop. Inside the floating aluminum lacework were seventy staterooms, a lavishly decorated staircase leading from one deck to another, a lounge with an aluminum piano, a dining room, and a bar. Meals were cooked on board by a continental chef, sumptuous servings of lobster, fowl, and roasts with any wine you could name. The four-hundred-dollar fare gave passengers a few unique days aloft. In fair weather the huge windows were left open, affording a panorama unmatched by today's commercial airliners. It was called the "Great Floating Palace."

Three days earlier, thirty-six passengers had boarded the 240-ton airship at Frankfurt and, except for head winds, the Atlantic crossing was uneventful. Nothing, exclaimed the *Hindenburg*'s passengers, could match travel by dirigible. The four twelve-hundred-horsepower diesel engines in their aft locations could barely be heard. Flight was vibrationless. A famous passenger, coasting along at eighty-four miles an hour, once poetically ventured that the sky giant was "held aloft by angels."

The huge transport was said to be the safest aircraft ever built, even though its sixteen balloon-cloth gas cells held seven million cubic feet of hydrogen, one of the hottest-burning gases known. An article in the current issue of *Collier*'s had said, "Only a stroke of war or an unfathomable Act of God will ever mar this German dirigible's safety record." So stringent were fire precautions that Lloyd's of London had insured it for five hundred thousand pounds at the low figure of five percent. The low-vapor fuel oil was so safe it could quench fire. Every one of the vigilant sixty-one-man crew was trained to spot hazards. Heinrich Kubis, the chief steward, took custody of a sparking windup toy that the Doehner children ran across the deck. "We don't tempt hydrogen," he said firmly.

Each crewman wore antistatic asbestos coveralls and hemp-soled shoes as he moved about his duties inside the cavernous skeleton. Ladders and catwalks were rubber covered. All matches and lighters were confiscated on boarding, and below the hydrogen bags was a specially insulated smoking room with a double-door entrance. This room was pressurized to keep stray hydrogen out, and a steward was constantly in attendance to light the passenger's cigars and cigarettes and to make certain no fire left the room.

The *Hindenburg* was running ten hours late. It had been due to moor at Lakehurst Naval Air Station at eight A.M., but it did not arrive over Boston until eleven-forty. Then, ominous black clouds over New Jersey prompted Max Pruss, the airship's captain, to hold out for better landing conditions. It was raining at Lakehurst when the airship was first sighted through the gloom at four o'clock in the afternoon. Lightning flashed in the nearby storm clouds. Veteran airship commander Charles Rosendahl radioed Pruss that he approved of his decision to wait. The *Hindenburg* droned low over the crowd of waiting friends and relatives, and its hoarse, throbbing engines added to the awesome sight. It swung over New York City to wait out better landing conditions.

With Captain Pruss in the control gondola was Captain Ernst Lehmann, a senior airship officer who had commanded the LZ-98 during World War I and who was the *Hindenburg*'s captain the year before. Both Pruss and Lehmann had commanded the *Graf Zeppelin*. Lehmann was along to advise Pruss on his first command crossing with the newer airship. His son had died only a few days earlier and he had not really wanted to come, but he thought he might have an opportunity to talk with officials in Washington about helium. The new and larger *Graf Zeppelin II* was under construction in Germany, and the Germans wanted and needed this rare gas that would not burn and that was commercially manufactured only in the United States. Lehmann probably knew what Washington's answer would be; they had no intention of giving Hitler a military advantage.

Over New York City, the *Hindenburg* circled the Empire State Building and headed south again, where a wet disgruntled crowd waited impatiently. One bored newsreel photographer went to a movie in Toms River; another reporter waited in a New York bar. Both missed the greatest aviation disaster of the thirties.

It was raining at the air station at six-twelve. The wind steadied at eight knots. The ceiling was still low, two hundred feet, but the visibility had widened to five miles. Rosendahl radioed the *Hindenburg,* "All clear and waiting."

Still, Pruss was cautious. He held off. At seven o'clock, with darkness coming on, Rosendahl recommended an immediate landing. The rain slackened to a drizzle, the wind died to two knots and the ceiling lifted to twenty-five hundred feet. The milling spectators, newsmen, and custom officials became even more restless and irritable with the approaching darkness. Many had been roused from their beds twelve hours earlier, expecting the landing on schedule. A landing crew of ninety-two Navy men and 138 civilian "dollar detail" volunteers huddled in the drizzle as the *Hindenburg* flew overhead, made a tight turn, and headed into the wind.

Pruss ordered hydrogen valved at seven-nineteen. The big ship weighed off and began to settle. Then Pruss called for water ballast to be jettisoned, but the airship was still slightly tail heavy. Six men were sent to the bow to trim it and the *Hindenberg* smoothly steadied its descent in the cool, wet air and leveled off under perfect control. Lehmann watched approvingly. Pruss, he knew, could handle the giant during this exacting and final part of the flight.

At 360 feet, the *Hindenburg* weighed off again and nudged onto the field. Seven hundred feet from the mooring mast, Pruss ordered the engines reversed to "idle astern," and the dirigible drifted slowly to a near standstill. Spectators and passengers shouted and waved, but one of the crowd, a former airshipman, thought there was "something wrong" with the airship's approach.

At seven twenty-one the starboard-bow line was dropped two hundred feet to the waiting ground party and the port line followed. They landed just inside the mooring circle and Lieutenant Raymond Tyler, chief mooring officer, saw dust puff up from the dry rope coils as they struck the wet sand. He began to direct the landing operation as several members of the ground crew noticed a peculiar flutter of the outer fabric at the rear of the airship.

At that precise moment, one spectator was about to draw his companion's attention to what he thought was a small spark flickering above and below the tail assembly.

Then it happened.

Inside the structure two crewmen, Sauter and Lau, were the first to see it. They heard a *pop* and a *whuff,* and looked up as Number Four cell near the tail broke into a brilliant flash of red, blue, and yellow light. At that moment observers on the ground saw a deep-red glow in the belly of the huge silver monster. A split second later the crowd froze in horror. Rosendahl gasped as he saw flames spurt up just forward of the top fin. "I knew the ship was doomed," he said later.

Most of the passengers heard and felt nothing at first. In the control room Pruss and Lehmann felt only a mild shock. "What is it?" Pruss asked, thinking a landing line had snapped. Then he looked aft, saw the rosy glow at the tail, saw the ground crew scatter, and knew.

The fire rushed quickly forward through the superstructure to burst one gas bag after another and to feed the already white-hot inferno. The muffled detonations were heard fifteen miles away at Point Pleasant.

As the landing party scattered and the crowd fell back, the *Hindenburg*'s nose lifted to five hundred feet. Passengers on the promenade deck were tumbled by the lurch. Relatives on the ground screamed. They were certain no one would survive the fire, even as they saw passengers hurl themselves through the huge windows, plummeting to the ground, some to die, some to live.

Pruss's first instinct was to valve the gas in the forward cells, but wisely he decided to let the burnings stern fall first.

The bust of Field Marshal von Hindenburg toppled from its pedestal, and passengers grabbed for whatever supports they could reach. Over the pandemonium and hysterical screaming, Mrs. Doehner kept her head and saved two of her four children by throwing them out the window.

John Pannes and his wife had become separated during the airship's approach. Although they were not far apart when the fire broke out, he left his window—through which he could have jumped—to join her. Both perished.

Margaret Mather reached the boarding ramp as the flames collapsed its supports, and stepped down to the ground practically unscathed.

A cabin boy looked up to see the fiery wreckage descending on him, then a water-ballast tank let go, soaked him, and spared his life.

James O'Laughlin, who had survived an earlier airplane crash, later said of the airship's death throes, "I never flew in a craft that traveled through the air as easily. Even in breaking up, the *Hindenburg* was gentle to its passengers—those who lived."

Miraculously, during the thirty-two seconds that elapsed since the explosion, some did survive in the melting, white-hot framework that crashed to the ground. Rescue men from the landing crew ran repeatedly into the twisted, glowing wreck. Max Pruss dashed clear at the last moment, then rushed back again and again to help his crewmen. Badly burned and dazed, he had to be forcibly restrained from reentering the glowing wreckage.

Captain Lehmann did not fare so well. His clothes had caught fire and he was badly seared over much of his body. He stumbled back and forth mumbling, "I don't understand . . . I don't understand. . . ." At Paul Kimball Hospital he was placed face down on a table, bare from the waist up. Thoughtful, polite, and fully composed, though in great pain, he realized how badly he was burned. Passenger Leonard Adelt paused in the hallway and stepped inside the room.

"What caused it?" he asked.

Painfully, Lehmann looked up. He did not answer right away. Finally, he said, "Lightning." Their eyes met and in that brief moment told each other that this was not really the cause.

Word of the disaster traveled quickly around the world. In the early morning hours Dr. Hugo Eckner, director of the Zeppelin Company, was roused from his bed by a telephone call. It was Berlin correspondent Weyer. Eckner was drowsy and annoyed. "Yes," he snapped, "what is it?"

"I thought it my duty to inform you of some bad news we have just received from New York," Weyer said.

"Yes?"

"The *Hindenburg* exploded at Lakehurst and crashed in flames."

"No!... No!... It isn't possible!" Eckner stammered.

Weyer pressed him for a statement. "Do you think there is a possibility of sabotage, Doctor?"

"If it was in the air ...," the dazed airship pioneer faltered, "... then it must have been sabotage."

The following day Commander Rosendahl visited Captain Lehmann. The airship veteran lay motionless on his bed. He never complained, and he knew he did not have much longer to live. As carefully as possible, the two men reviewed every possible cause of the fire. Each one—from a static spark to a gas cell ruptured by a shattered propeller blade—led them in circles. "No ... no," moaned Lehmann, shaking his head. "It must have been a *Hollenmaschine* [an infernal machine]." Then he added hopefully, "But, of course, no matter what the cause, the next airship *must* have helium." He died later that afternoon without realizing there would never be a "next" airship.

The death toll stood at twenty-two crewmen, thirteen passengers, and one ground crewman. On May eleventh, ten thousand spectators and mourners gathered at Pier 86 in New York to witness the services for the twenty-eight European victims. Germany declared two weeks of official mourning, and as the SS *Hamburg* steamed out of port, the ghost of the airship era sailed with it. The hope of the rigid airship was laid to rest with the *Hindenburg*'s victims.

Two unnatural occurrences found their way into the *Hindenburg* picture. The first happened exactly one month to the day before the airship went to its catastrophic destruction. In Milwaukee, a German citizen had a frightening dream. In it he saw a dirigible drifting toward the mooring mast at Lakehurst. Then it burst into flames so suddenly and so completely that the man was sure all aboard had perished. The dream preyed on his mind and seemed so real he felt compelled to warn someone of the approaching disaster. Because he was an illegal alien who'd jumped ship in 1928, he lived in constant fear of being discovered and deported. He asked his landlady to write to the German ambassador. She did.

To Dr. Hans Luther, the woman's "friendly warning" and assurance that this was "no joke" was but one of many letters he received warning of the destruction of either the *Graf Zeppelin* or the *Hindenburg*. There was something special about this one, however, so he sent it on to Captain Lehmann, who put it in his coat pocket. It was there when fire scorched his coat as he fled the burning airship. It went back to Germany with the remnants of his personal belongings. There, Max Pruss saw it a year later.

The second curious incident involved Lehmann's wife Marie, who was fearful over this season's first *Hindenburg* crossing. Grief-stricken over the recent loss of their son, Frau Lehmann would have preferred that her husband remain at home for a while. A clairvoyant in Vienna had recently told her that her husband would die in a burning dirigible.

It was strange that the bulk of the *Hindenburg* letters and phone calls received by Dr. Eckner and Dr. Luther did not warn the ship away from Rio, Frankfurt, or Friedrichshafen, but only from Lakehurst.

Even before the charred and twisted wreckage cooled, aeronautical experts and technicians flocked to Lakehurst to find the cause of the disaster. Investigators arrived during the night and began to question witnesses. One young line

THE GREAT FLOATING PALACE 251

handler said he watched a boy and a dog jump from the front of the dirigible as it settled. They landed safely, he said, and ran into the pinewoods.

The Navy Department assembled a board of inquiry to convene the following Saturday, but the group of Navy Officers adjourned when the Department of Commerce reminded them that this was a civilian air disaster, not a military one. On Monday, a civilian board of the Bureau of Air Commerce met in an empty hangar of the air station. It was the same huge, drafty room that had been a waiting room and, more recently, a morgue for the airship's victims.

Rosendahl, gruff and outspoken, was the first to testify. He stated that when the *Hindenburg*'s bow lines hit the ground, they had "definitely grounded the airship" by discharging whatever static electricity it held. However, in later testimony, Lieutenant Tyler seemingly contradicted the commander when he claimed the trailing ropes could not have discharged any static electricity.

The controversy began from the first day. Nothing was firmly established then or in the days of inquiry that followed. The evidence accumulated, reams of it. And unknown to the board at that time, intrigue to match the enigma of the *Hindenburg*'s destruction was being brewed in Berlin. Hermann Goering, Germany's number two man, sent orders to the airship officers that there was to be no speculation as to the cause; they "should not try to find an explanation." Obviously, the death of the *Hindenburg* was a blow to the pride of Nazi Germany. If it were ever established that the airship had been destroyed by an enemy faction, the disgrace would have been unbearable. Thus the crew, many of whom feared Nazi power, complied. The conspiracy of suppression muzzled them. The commission, unaware of Goering's orders, did not learn until much later what the airshipmen suspected about the cause.

Electrical malfunctions were logically ruled out. Definitely, there had been no short circuits or overloads. Major structural failure was also ruled out; no lines or brace wires had snapped. A sticking gas valve? One had stuck on the South American run the previous season. Sparks from the engines? They were under the airship and well away from the buoyant hydrogen that would have risen far above them. An incendiary bullet? Bullets fired in tests failed to ignite identical gas cells. Besides, no one knew in advance which approach the *Hindenburg* would use when it came onto the mooring mast. Static electricity? Perhaps, but how, or why, would a static spark travel from the bow lines all the way to the tail before causing an explosion? Some technical reports deny that St. Elmo's fire and static electricity are strong enough to ignite even sensitive hydrogen. Dr. Eckner ruled out ball lightning and exhaust sparks. A broken propeller blade? No evidence.

What about a radio bomb, a device smuggled aboard with a single frequency matched to a transmitter that had been triggered by an accomplice watching the landing from the pinewoods? Highly unlikely for many reasons. The tiny transistor and the printed circuit were electronic dreams in 1937; such a bomb in that day would have been too large and bulky to escape detection. Something of an electrical nature was found, however. Something that was not seriously considered at the time, despite its having come to the attention of the board. A small, unidentified part of the wreckage came under the scrutiny of Detective George McCartney of the New York City Police Bomb Squad. He had it analyzed. It proved to be a cotton container of manganese dioxide, zinc oxide, and graphite—the elements of a dry battery. No one, crew or passenger, was permitted to have such an item in his possession. Gestapo agents in Frankfurt had thoroughly searched all passengers' belongings as well as the airship before takeoff.

One other strange finding was made in the wreckage, something that probably belonged to one of the *Luftwaffe* officers who was carried as a passenger. It was a Luger pistol. One shot had been fired.

On the stand, the Germans who were able to testify were

cautious and reticent. Captain Witteman, one of the five qualified captains aboard, had emerged from the incandescent blob of fire practically unscathed. He had rushed up to Rosendahl soon after the crash and told him he wanted to talk privately with him as soon as possible. But to the board he simply said, "The whole affair is a complete mystery to me."

At one point in the proceedings it was suggested that, during the tight final turn toward the mast, a brace wire had snapped and whipped into cell Number Four, where leaking gas was ignited by static electricity. Chief Engineer Sauter and Helmsman Lau, who were in the vicinity of the initial explosion, both said no.

In Lennox Hill Hospital in New York, *Luftwaffe* Major Frank Witt could not appear to testify, but from his bed he told of threats and an anonymous letter his brother and a representative of the Zeppelin Company knew about.

Hans Freund, a rigger in the tail section at the time of the explosion, had to be coaxed to talk. He was recalled for questioning the following Monday. Yes, he said, gas valves could stick. No, he did not believe hydrogen was leaking. He also established that Ludwig Knorr, the chief rigger who was killed, had relieved another rigger, Eric Spehl, at Number Four gas cell at six P.M. one hour and twenty-one minutes before the explosion. This information later proved helpful to the Gestapo, although it was inconclusive.

Steward Nunnenmacher took the stand, as tight-lipped as his predecessors. He said nothing of a distraught passenger's unusual behavior in the dining saloon during the landing delay. The *Hindenburg* was maneuvering about the air station, preparing for the approach, when the man began to pace nervously back and forth near an opened window. This was twenty minutes before the explosion.

"I don't want to go around the field again," he mumbled. "I want to go down!"

Nunnenmacher made an effort to calm the passenger by explaining that the airship had to await better conditions.

When Chief Steward Kubis entered the room a few minutes later, the man was even more upset.

"My watch," he cried out, "my watch! Can I find it when we go down?"

In his nervous state he had removed his wristwatch and dropped it over the side. Nunnenmacher and Kubis knew no reason for the man's hurry.

When Kubis took the stand, he was questioned at length but, like Nunnenmacher, he said nothing of the jittery passenger. Later, the FBI investigated and found the man to be passing himself off as an American citizen, although he was traveling on a foreign passport.

Kubis also omitted another important point of which he had full knowledge. It concerned Rigger Knorr, who at seven o'clock looked over Number Four cell and thought it somewhat light of gas. A leak perhaps? Perhaps not. They had reduced altitude and this might account for it. Knorr decided to look at the other cells for comparison. As he worked his way forward along the catwalks and labyrinth of ladders, he passed Kubis's door. He paused, mentioned his discovery, and suggested that some quick repairs might have to be made before the turn-around trip. Kubis must have thought it strange that a leak had occurred so quickly, but he said nothing about it to the board.

As the possibilities were examined and considered, the theory of sabotage persisted. The board was anxious to investigate this avenue thoroughly. They failed, and to this day the deliberate destruction of the *Hindenburg* has never been proved, disproved, or eliminated.

The board adjourned on May twenty-eighth. Their official finding: "most probably" St. Elmo's fire, a type of static electricity. Forty-four years after the accident, with the fear of Gestapo reprisal gone, the scraps of information given by the surviving passengers and crewmen reject this finding and confound the mystery even more. And still the truth has not come out.

The dirigible's destruction by sabotage was not unlikely.

Look at the record. Time bombs were planted on two postwar passenger zeppelins, the *Nordstern* and the *Bodensee*. Another was found under a chair in the lounge of the *Graf Zeppelin*. America's own *Akron* was involved in a sabotage plot in 1931, but the damage was discovered and repaired before flight. When the *Hindenburg* was on its South American schedule the year before, this plot was uncovered; a passenger's strange behavior had attracted the Gestapo's attention. While the man was absent from his Frankfurt hotel room, the agents found detailed interior sketches of the *Hindenburg* and *Graf Zeppelin* in his belongings. There were several Xs marked at the top of the gas cells and along critical places in the fuel lines. Somehow the suspect got warning of the search and escaped.

Late in 1938, the Hamburg-American liners *Reliance* and *Deutschland* were also involved in unexplained fires.

Dr. Ludwig Duerr, designer of the *Hindenburg,* was absolutely convinced of sabotage of the dirigible.

Who had reason to destroy the giant airship? A crank? A pyromaniac? A member of the *Gegen-Nazi Weiderstand*— the anti-Nazi resistance? How much did the Gestapo know of the plot beforehand? Years after the disaster, when the German-Russian military pact was made, the Gestapo quietly confided to Max Pruss that their search had led them quite close to the Communists. Pruss himself had long suspected a certain passenger, but although the Gestapo never uncovered more than circumstantial evidence, they did narrow their investigation down to one man, rigger Erich Spehl.

No one knew much about the tall, fair-haired crewman. He was a moody and introverted young man, aloof and quiet. He did his job well, bothered no one, and occupied his spare time with photography. Shortly before the *Hindenburg*'s scheduled departure, Spehl was seen hanging around one of the drinking places frequented by Communists. He was keeping company with a somewhat plump woman several years his senior. His crewmates remarked to one another that this behavior was quite unlike him, but Spehl, obviously enamored of the woman, explained that it was just a farewell party.

Little else is known of the mysterious woman except this: As the *Hindenburg* drummed over the Atlantic, she visited the Frankfurt terminal of the Zeppelin Company three times. She was clearly apprehensive and highly curious. Each time she inquired about the airship's position. On her last visit she wanted to know the hour of its landing at Lakehurst.

Spehl's position near the tail was in the area where the explosion occurred. He had been relieved by Knorr, and at the time of the landing approach was in the farthermost part of the nose. He perished in the holocaust, so we shall never know if he played a part in setting a fire device in the crevices of gas cell Number Four. Purely circumstantial evidence suggests that he may have hidden a flashbulb filled with aluminum foil, near the center catwalk under the aft bag. It would have to have been connected to an electrical source (the small battery?) and a timing device. Here is what supports this theory:

1. Spehl associated with known Communists who were active in trying to destroy the Nazi myth of invincibility.
2. He had full access to the cavernous interior of the airship, much opportunity to work alone and undetected, and could have placed the device in position before he was relieved of his watch at six P.M.
3. With his knowledge of photography, he knew the potential of flashbulbs, that they contained their own oxygen, could instantly reach sixty-four hundred degrees Fahrenheit, and could be made to explode when fired by simply nicking the glass with a small file.
4. He was among the few crewmen who were farthest from the explosion when it occurred.
5. Crewmen in the tail aft of cell Number Four said the fire started as a brilliant flash (like a photoflash bulb),

and sounded like a *pop* (as when a bulb is accidentally shattered because of faulty construction).

True, it is all circumstantial. And it seems to lose significance when we remember that the airship was scheduled to land at eight A.M., and its destruction occurred twelve hours later. Logically, a bomb would be timed to explode over the Atlantic where its quick work would leave no trace of the crime.

We shall probably never know for certain who, or what, destroyed the *Hindenburg*. Or why or how. The disaster could not have happened had helium been used. In the enigmatic ending of the short but brilliant day of the giant rigid airship, the dirigible had shown its unfitness to everyone but the most devoted lighter-than-air enthusiasts. The insurance companies paid the Zeppelin Company two and a half million dollars, with the announcement that never again would they underwrite a hydrogen-filled airship.

The *Graf Zeppelin* was near the Canaries on its return run from Brazil when Captain Schiller received the news of the *Hindenburg*'s destruction. After he arrived at Frankfurt his airship was put into its shed and never flown again.

Work on the *Graf Zeppelin II* continued at Friedrichschafen, where it was tested in September of 1938. But Hitler forbade even one commercial flight with the new giant. This was his revenge for Dr. Eckener's earlier refusal of the Friedrichschafen hangar for a Nazi mass meeting. Work on another superzeppelin, the LZ-131, was halted at the outbreak of the war. Then, in March of 1940, after bitter words between Max Pruss and Reichmarschal Goering, both *Grafs* were dismantled. Two months later, on May 6, 1940, the massive airship shed at Frankfurt was blown up. It happened three years to the day after Lakehurst.

Like Rosendahl, Pruss never lost faith in the dirigible as the best means of air transportation. In 1957, he journeyed to Brazil and the United States, trying to raise capital for another airship company. He was unable to arouse enough interest and the dream never materialized. Pruss died in 1960, still satisfied that the cause of the *Hindenburg*'s destruction was sabotage.

Traditionally, the blame for the disaster has rested on the zeppelin's lifting force—the sensitive and unstable hydrogen in its sixteen gas cells. But recently, fuel and energy technicians have risen to the defense of hydrogen as a safe fuel, and brought forth one of its properties that had been overlooked in the past. Hydrogen, they say, is so light that, when ignited and burned, the flames move quickly *upward* and *away*. To prove their point they use the very motion picture film of the burning *Hindenburg* and, in slow motion, show how the burning hydrogen did indeed move away from the people in the gondola, promenade deck, and passenger cabins. What, then, caused the deaths of the crewmen and passengers in these areas? The exploding diesel fuel tanks, say the technicians. In bursting, they sprayed the liquid fuel along the bottom of the burning ship.

The thirteen passengers who died aboard the *Hindenburg* were the first and last commercial casualties in airship history. Not a single rigid airship has carried another paying passenger since May 6, 1937.

In the words of Dr. Eckener when he discussed the *Hindenburg*'s end with newsmen in Berlin, "It is over."

14

"Do You Want to Live Forever?"

For hundreds of fans of a famous bandleader, his unforgettable music, and his determination to bring a touch of home to weary American soldiers, one flying mystery remains an enigma of World War II. The circumstances of the musician's strange disappearance on the day before the Battle of the Bulge remains as obscure as the fog that swallowed his single-engine airplane on December 15, 1944. What happened to Major Glenn Miller after he took off from a small country airfield in England to entertain GIs in Paris is still unknown.

From 1938 to 1943, the tall, serious musician with the rimless glasses and trombone built a band that rose to the peak of popularity in America. He did it through a new sound that he'd been searching for—a blending of clarinets and saxophones that gave his group a distinctive sound. Its matchless style caught the pulse of the public and never let it go; it lives among the classic favorites of today. Glenn Miller was part of the big band era. His records "Moonlight Serenade," "String of Pearls," "Chattanooga Choo-Choo," and "Tuxedo Junction" outsold all others. With "In the Mood" as his identifying theme, he found a place in the hearts of swing-era Americans that was rarely equaled in popular music.

The band's musical quality was far from accidental, for Miller was a disciplinarian when it came to performance. He also had a different effect on everyone with whom he came into contact. Some found him complex and somber, cold, interested only in commercial gain. Others found him warm and approachable. A few who knew him well, however, agreed that he harbored a trace of fatalistic melancholia in his sensitive nature.

Whatever drove him, he always seemed to know what he wanted to do and where he was going. No one, then, was greatly surprised when he applied for a commission in the U.S. Army in 1942. The fact that he was overage—he was thirty-nine—and exempt from military service because of poor eyesight did not interfere with his plan to take his music to the fighting men at the battlefronts. He left his wife Helen and their adopted son (they were to have an adopted daughter later) and formed a new sixty-man Army band.

Captain Glenn Miller of the Eighth Air Force took his new band of the American Expeditionary Force to London just after D-Day. The V-1 buzz bombs were striking hard at the city, so quarters were found for the musicians at Bedford, a town about fifty miles north. From studios there, they began a grueling schedule of personal performances and recordings, sharing the facilities with the British Broadcasting Company's Symphony Orchestra. It was here the bandleader came to know Cecil Madden, the BBC producer. They became good friends.

One day, at Miller's apartment on Waterloo Road, he and Madden were enjoying a few quiet hours before the next performance. Miller showed the producer a scale model of a house he planned to build in Monrovia, California, after the war. He had named it "Tuxedo Junction," after one of his most popular recordings. It was a mansion suitable for a Hollywood star, but Glenn seemed strangely morose about the future. "It's nice, of course," he told Madden, "but somehow I don't think I'll ever get back to California." The comment puzzled Madden. Why would a noncombatant officer, a bandleader, have such a premonition? In London, Miller made a similar comment to Lieutenant Don Haynes, the administrative manager of his band. "Don," he said, "I have a strong feeling I'm not going to see Helen and Stevie again. I know it sounds odd, but I've had the feeling for some time. . . . The Miller luck has been phenomenal for the past five years, and I don't want to be around when it changes.'' Haynes later said that Miller believed something was going to happen to him—that possibly one of the buzz bombs was marked for him.

The band schedule continued at a punishing pace and began to take its toll. There were appearances at bases throughout the island and the band traveled in two-engine C-47s. Miller had a dread of flying and the frequent trips in sometimes marginal weather left him with a knotted stomach and frayed nerves. But the strict upbringing of the tall midwesterner had conditioned him to hide his fears.

Despite the pressures, Miller, now a major, continued to push his plan to take his music to the men at the front. He was already playing to the steady stream of soldiers arriving from America enroute to France to expand the invasion bridgehead. Paris was the Allies' prime objective, and during the summer of 1944 Miller told the troops, "When you fellows get to Paris, I'll play you a special Christmas show there." They liked that, and Miller began to get the paperwork rolling to make it come true. Finally, on November fifteenth, he was summoned to Supreme Headquarters Allied Expeditionary Force at Versailles, outside Paris. A staff officer asked him, "How would you like to bring your band over here next month and entertain the GIs and the troops in the hospitals? Possibly stay for six weeks?"

"Great!" Miller said. "Just what we wanted to do!"

There was a problem, however, and Miller thought about it on the flight back to England. His prime responsibility— the reason he had in fact been brought to England by General Eisenhower—was to play primarily for a radio audience that numbered in the millions. There could be no break in that schedule, he knew. In London he talked with Maurice Gorham of the BBC about a plan to have the scheduled radio shows *and* go to Paris as well. "If we recorded six weeks of programs, we could do it," he told Gorham. The producer pondered the idea. "That's a lot of work," he said, "but if you can get six weeks in the can before you leave, yes, it can be done."

Back in Bedford, Miller assembled the band and put the challenge to his musicians. Everyone accepted it. The next day they began a series of exhausting recording sessions to build a backlog of programs so there would be no break in their radio shows for the troops. This was in addition to their regular schedule of live broadcasts. The different band units were already doing seventeen shows a week, but to a man they all set to work without stopping. For some sessions they arose early in the morning and played until the early hours of the following day. Only disciplined musicians, led by a dedicated and determined conductor, could record eighty-five half-hour shows in an eighteen-day period.

The growing pressures were evident in Miller as well as in several of his musicians. The strain of the buzz bombs, his dread of flying, and the lingering doubt that the band might not be able to meet the Paris schedule for the Christmas show he'd promised the troops, preyed on him. He ate poorly and his periods of depression became more evident. His health was not good and he had lost so much weight that his once well-tailored uniforms now hung loosely on his lanky frame. Nervous and restless, he often let go in severe reprimands to the band members as he pressed on with the recording marathon. Then he would calm down, plunge back into the schedule. When the final recording had been made on December twelfth, he told Gorham at the BBC and then immediately prepared to airlift his entire band to Paris. Only one program had not been recorded: the Christmas program, which he would conduct from Paris.

Meanwhile, Don Haynes had flown to Paris to find quarters for the band. While there, he met Colonel Norman Baesell, the executive officer of the Milton Ernest Post just outside Bedford. Baesell was well known to Haynes, Miller, and the band, and he made frequent trips to Paris. He was also an officer with military contacts and was able to bend regulations on occasion. When Haynes returned to London he reported that facilities were ready for the band. Major Miller's orders, issued by command of General Eisenhower, directed him to leave England on or about December 14, 1944, by Air Transport Command aircraft. Haynes had been originally selected to precede the band by a day or two, but Miller, still impatient, decided to go instead. On December thirteenth, Haynes drove Miller to the London ATC terminal they used for Paris departures—Bovingdon—but the fog was so thick that no SHAEF shuttle planes could fly. Haynes left Miller in London to await clearing weather and drove back to Bedford. According to Haynes, who allegedly kept a "war diary" and who died in 1971, here is what happened next:

The following day—the fourteenth—Haynes again ran into Colonel Baesell, who told him he was flying to Paris the next day in General Goodrich's utility plane. Together they agreed that Baesell could take Miller along when the general's plane landed at a small RAF training field at Twinwood Farm, four miles north of Bedford. Haynes telephoned Glenn in London, and the bandleader agreed to ride along as a passenger the next day. Haynes drove down to London and picked Miller up at the Mount Royal Hotel. They returned to Bedford in time to have dinner with Colonel Baesell at the officer's club after which the colonel excused himself. Miller and Haynes sat up until three-thirty A.M., talking about plans for the band after the war. Outside, the weather continued to be dismally wet and cold.

On the morning of the fifteenth, a Friday, Baesell telephoned Miller at about nine A.M. to say that the weather was still bad but showed signs of clearing by early afternoon. He suggested they all meet at the officers' club for lunch. Glenn and Don read the papers and relaxed in the lounge to kill time. Then, still restless, they walked to the band members' quarters to make sure everything was in order for their flight across the Channel on the next day.

During lunch with Baesell, the colonel was called to the telephone. He returned and told them his pilot, Flight Officer John R. S. Morgan, had just received clearance at Station 595, the 35th Depot Repair Squadron to the northeast at Abbots Ripton, and would pick them up at Twinwood Farm within the hour. Outside, the rain was still falling. Miller looked dubious.

En route to the airfield, with Don Haynes driving the staff car, they paused outside General Goodrich's chateau while Baesell ran in to speak with the bedridden officer for any final messages or instructions. Dark clouds hung low in the sky and the rain increased. "I don't think the pilot will ever be able to find the airfield," Miller muttered. Colonel Baesell ran out of the chateau, got in the car, and a few minutes later they pulled up beside the Twinwood Farm flight control tower. Haynes shut off the engine and they sat quietly in the car, waiting. The rain, now reduced to a drizzle, splattered against the windows and drummed against the car's roof. Low stratus clouds still hovered overhead, scarcely two hundred feet from the ground. Haynes said later that, about that time, he too was beginning to doubt that Flight Officer Morgan could find the field.

They talked and smoked for about thirty minutes, then Baesell left the car and climbed the ladder to the control tower. He was back in ten minutes.

"I talked with the station," he said, "and they told me that Morgan left there in a C-64 fifty-five minutes ago, so he's due here any minute."

Miller was visibly nervous, Haynes said, and he got out of the car. He looked up at the sky. Haynes and Baesell also left the car and all three walked over to the control tower. It was miserably wet and cold. Haynes looked at the thermometer on the building. It registered thirty-four degrees Fahrenheit. As they walked back to the car, Miller again said that Morgan wouldn't find the field. Then, out of the north, they heard the faint, steady drone of an aircraft engine. Baesell smiled. "There he is. It's a single-engine plane and it sounds like a flock of outboard motorboats." The flippant comment must have been disquieting to Miller, who was already edgy about another flight in marginal weather. Then, as the engine grew louder, it sounded as though the plane was flying directly over the field; but in the overcast they were unable to see it. As the plane headed south, the engine noise faded. Miller turned to Baesell. "Like I said, Colonel, in this weather he couldn't even find the field; he missed it."

Baesell shook his head confidently. "Don't count on it, Glenn. Morgan's flown thirty-two missions in B-24s and he's used to this kind of weather. I say he'll be on this strip inside of ten minutes." As Baesell spoke, the engine sound changed pitch, which meant the plane was turning, and a few minutes later it burst out of the two-hundred-foot overcast directly over the field and made a low-altitude circle. The tower had talked Morgan in.

The plane was a Noorduyn Norseman, a high-wing monoplane with a 550-horsepower Pratt and Whitney radial engine. Aside from the pilot and copilot, it could carry eight passengers and was a popular plane for transporting staff officers on short trips. The Norseman was a sturdy and reliable airplane; it cruised at 150 miles an hour and had a range of six hundred miles with 125 gallons of fuel.

F. O. Morgan throttled back and rolled the plane into a gliding turn upwind, then touched down on the hard-surface runway. The three officers drove out to meet the plane and pulled up alongside of it. Morgan kept the engine running as the men stepped from the car. He opened the window and waved. "I'm sorry to be late," he said. "Flew into some heavy squalls, but the weather's supposed to be clearing over the Continent."

Colonel Baesell handed Morgan his bag and went back to the car for another package. Haynes tossed Miller's B-4 bag into the passenger compartment and shook hands with Morgan. Baesell climbed aboard and seated himself beside Morgan in the copilot's seat. Miller seemed to hesitate at the door. "This plane has only one motor," he said to Baesell.

The colonel replied, "What the hell, Lindbergh only had one motor and he flew clear across the Atlantic. We're only flying to Paris."

Miller stepped aboard and sat in a fold-down bucket seat directly behind Baesell. It faced the opposite side of the plane. Morgan sat down in the pilot's seat and, as they fastened their seat belts, Haynes held onto the still-open door, pushed against him by the slipstream of the idling engine. He waved his hand at Miller. "Happy landings and good luck," he said. "I'll see you in Paris tomorrow."

Miller replied, "Thanks, Haynsie; we may need it." Then, before he closed the door, Haynes saw Miller look quickly around the cabin compartment. He turned to Baesell and asked over the colonel's shoulder: "Hey, where the hell are the parachutes?" Baesell replied jokingly: "What's the matter, Miller, do you want to live forever?"

At this point, Don Haynes said, he closed the door, secured the latch and stepped back to avoid the propeller blast. Morgan waved to him and opened the throttle. At one fifty-five they started down the runway with a roar, gained speed, and slipped into the cold, wet air. In a few seconds the plane's indistinct shape disappeared into the overcast and a few minutes later the engine noise faded away to the south. Haynes said he got into the car and drove back to Bedford. In the postwar years several researchers found reason to question several aspects of Don Haynes's account.

The Norseman was to have traveled on the outbound route for Bordeaux via A-42—Villacoublay—near Versailles, which was the Army Air Force's designated route for transport aircraft flying to the southwest French coast. Morgan's intentions, however, were not to proceed to Bordeaux but to turn off the route and land at Villacoublay.

That evening no one—band members or the vast Glenn Miller following—had any reason to suspect anything was amiss with the American band of the AEF. At eight thirty, the program was heard as usual over the BBC. The same musical magic was there—perhaps even a trifle better. And

Glenn Miller's unmistakeable midwestern drawl made the usual announcements and introductions. It was, of course, the first of their many recordings.

The following morning, the sixteenth, the band drove by bus the fifty miles to RAF Bovingdon, outside London, to board the three waiting C-47s that would fly them to Paris to join Miller. The weather was bad and, after waiting a few hours, the men returned to Bedford. The next morning they drove again to Bovingdon, and again they were turned back because of bad weather.

On the morning of the eighteenth, however, the sky cleared and, instead of traveling again to Bovingdon, the C-47s flew to Twinwood Farm, picked up the musicians and flew directly to Orly Field, outside Paris.

Oddly, they did not see the major's familiar face as their planes taxied to a stop and the engines were shut down. Haynes left the group for the operations office to phone Miller and find transportation to their hotel. Two hours later Haynes returned with two buses. As the men boarded, they asked Haynes, "Where's the major?" Haynes replied that he didn't know—no one seemed to know—but when they got settled into their quarters he would get in touch with him.

Haynes's queries the next day uncovered nothing as to the major's whereabouts. He cautioned the band members not to write home that Miller was missing. He checked at the Eighth Air Force Headquarters and was told that all flights out of England had been grounded for the past week—except for a report of one flight bound for France. It was single-engine plane, but it had not been heard from. Next he checked the hotels and restaurants he knew Colonel Baesell frequented, but no one had seen him—or Major Miller. At SHAEF headquarters Haynes asked General Barker to telephone General Goodrich for news of Baesell, Morgan, and Miller. General Goodrich told Barker that Morgan and Baesell were due back but that neither had returned. He was furious when he learned that Morgan had tried to fly over the channel in a C-64 under such hazardous icing conditions. He said he would have a search made at daybreak. When General Barker hung up, he turned to Haynes and said, "It looks very bad, Lieutenant. I'm afraid Major Miller has had it."

It appeared that the plane must have encountered severe icing conditions and crashed somewhere—in England or in the channel—before it reached France. Oddly, no reports were released to indicate the plane was either seen or heard after it left Twinwood Farm, or that a search was made. The Battle of the Bulge began the day after the plane disappeared, which may account for all air resources being directed on missions elsewhere. Meanwhile, in Paris, Jerry Gray took over the baton, and rehearsals for the Christmas program began. The situation was strange; in the evenings the musicians huddled by candlelight in their unheated hotel rooms, and talked about the major's disappearance. Some held to the slim hope that his plane had merely been forced down in a remote area and he was out of touch with communications. Sometime before Christmas, Glenn Miller would appear and keep his promise to the GIs.

Every evening the London BBC recordings of Glenn Miller's band took the place of the live show with no one the wiser. It was pure deception, but there was no choice; the American Army was holding all casualty lists until after Christmas. The familiar Miller renditions were heard every evening: "American Patrol," "Pennsylvania 6-5000," and his ever popular "In the Mood." For the band members, who heard their leader's calm, recorded voice over their portable radios, and knowing they would never see him lift his baton again, those were miserable days. Then, as Christmas week drew near, a rumor circulated through the theater that Miller had been captured or had defected. As the time for the Christmas Day broadcast approached—the one for which the band had made no recording—the War Department was forced to release the news of Miller's disappearance. The announcement was made on Christmas Eve, after Helen Miller had been told in Tenafly, New Jersey.

In England, gloom settled over many in the Allied camps, in apartments and houses. A woman storekeeper at the Chilwell Ordnance Depot said: "We all wept when he was killed." Thousands of other women shed tears with that bleak announcement that carried Glenn Miller from orchestra fame into legend and mystery.

If anything feeds a rumor mill and promotes speculation it is the sudden disappearance of a famous person. The confirmation that Miller was now officially missing sparked a rash of theories about his death—several of which live to this day. Most of them were clearly third- and fourth-hand rumors, totally without basis. Miller was a secret agent and his disappearance was planned; or Miller was a black marketeer who had made a fortune in war goods and persuaded Morgan and Baesell to fly him to Germany or behind the Russian lines. One variation of this was that Baesell was involved in the black market, shot Miller and Morgan en route to Paris, landed the plane himself, and went into hiding after the war. Another report said that Miller was shot down and was physically and mentally incapacitated—or horribly disfigured—and is living now in a secluded convalescent home in the Midwest. All of these speculations were eventually shown to have no basis in fact, but other theories have not died as easily.

Historians, writers, and Miller buffs—especially those of the Glenn Miller Society with headquarters in Dorset, England—would still like to know, after three and a half decades, what became of the single-engine Norseman as it headed south from Twinwood Farm. The members work to re-create the Miller mood at their gatherings and schedule the few remaining musicians of Miller's original band for their frequent recitals. George T. Simon, a former member of Miller's stateside band, said in his biography *Glenn Miller and His Orchestra,* "Why should it be so difficult to accept the most obvious: that Glenn Miller was simply the victim of some terribly bad judgement on the part of a colonel, [who was] too anxious for his own good to get to Paris, and a pilot, too adventuresome for his or anybody else's good." While Simon's point is valid, and goes a long way toward dispelling the many ridiculous rumors, the two questions foremost in the Glenn Miller saga remain unanswered: *Where did he crash?* and *Why?* There are, today, three promising areas for study: one, the investigation of John Edwards; two, a case of mistaken identity; and three, the account of former WAAF Leading Aircraftwoman Muriel Dixon.

I first heard of John Edwards in 1974 when a UPI release said he claimed to have found the wreck of an aircraft off the English coast that could be the plane in which Miller disappeared. I put the story in my file and when I began to collect my research material for this chapter I decided it was time to locate Edwards and obtain first-hand information from him. After six months of tracking down leads, I found him in Nottingham, England.

The startling conclusions of John Edwards have been received with mixed feelings by the rank and file of the Glenn Miller Society. Edwards, an electronics company executive, is clean-cut, personable, and intelligent. He is an avid researcher who spent considerable time and money on the Miller mystery before he abandoned it, satisfied that he had solved the puzzle. In August of 1979 he told me about his research project that covered sixteen exciting years.

Edwards began his inquiry in July of 1959 and ended it in 1976, after spending fifteen thousand pounds on the investigation. Basically, Edwards said, it began when he was in a search-and-rescue squadron of the Royal Air Force. He ran a Forces broadcasting station—a record request show—and he used "In the Mood" as his signature. He liked Miller's music; then he heard of the mystery of the bandleader's disappearance. He knew the work of a search-and-rescue squadron was accurate in their reporting of missing aircraft and flying accidents. He said: "I was horrified to hear that—allegedly—no search was carried out after he vanished. So I vowed to begin an investigation into the loss of Glenn Miller and carry it through."

Edwards began by collecting information from official records. He got a copy of the missing aircrew report from the U.S. National Archives from which he learned many particulars of the last flight. Then he used the British Meteorological Office to research the weather conditions over a twenty-four-hour period for that day, at every airfield along Miller's route. Finally, he obtained an airplane similar in flight characteristics, speed, and icing tendencies, to the Norseman—a slightly modified Cherokee 140. His first flights were sweeping searches of the United Kingdom along the route during the winter when the trees were bare, to see whether the aircraft had crashed on land. "We found nothing," Edwards said, "and came to the conclusion that Miller went into the sea, which of course was a harder area to search."

Each winter Edwards waited at an airfield in England until the weather conditions were exactly the same as they were on Miller's flight. Then, with the specially equipped Cherokee, they flew under and through the overcast sky to collect data on airframe icing, which they believed was the cause of the crash. They found this theory was wrong, Edwards said. The cause of the crash wasn't airframe icing along the route and over the channel, he discovered, but carburetor icing.

Edwards contacted Pratt and Whitney, the American manufacturer of the engine, who sent him drawings and specifications. The technical data on the Model S3H1 engine showed a thirty-six-degree temperature drop through the carburetor during normal operation. He was able to establish that the outside air temperature was minus three degrees at a thousand feet at the time of Miller's plane crash. Edwards even traced the serial number of the plane's engine so positive identification could be made if the aircraft was found.

Now Edwards had to know *where* it crashed. He used the Honeywell Mark III time-sharing computer in Cleveland, Ohio, into which he fed the data he had gathered on the Norseman's flight path. The result was, according to Edwards, a most probable position printed out by the computer. Next he used—again American equipment—a device called side-scanner sonar. This was used during the survey for the British Channel Tunnel—the tunnel to connect England with the continent. Edwards was successful in having the survey people deviate from their course and take pictures for him with the special equipment. The result, Edwards claimed, was a picture of an aircraft on the sea bed. It had exactly the same configuration as the Norseman in which Miller reportedly disappeared. Thirty years to the day after Miller's last flight, Edwards made a commemorative flight from Twinwood Farm to the location in the channel. But now Edwards and his associates encountered a serious inconsistency: the timing was way off.

They discovered the error this way: To confirm Edwards theory that Miller's plane was in the right position as determined by the computer—about eight and a half miles off Dunkerque—the investigators sought out eyewitnesses. They found people from the German side as well as the British side who actually saw the plane crash. But the only computer data that didn't agree with the eyewitnesses' testimony was the *time* of the crash. There was a two-hour difference. The computer findings said the plane crashed around four P.M. in the afternoon; the eyewitnesses said they saw it crash about six P.M. in the evening. Edwards knew, however, that the Norseman didn't carry enough fuel to stay in the air that long; the three men should have landed in Paris well before six P.M. "I went back to the drawing board again," Edwards admitted.

Then Edwards learned of a retired USAF Lieutenant Colonel, Tom Corrigan, who also had an interest in the Miller story. Corrigan was able to clarify some misinformation relative to the Miller flight as well as bring up yet another—and dramatic—aspect of the major's disappearance. Corrigan knew it was generally accepted that all flights from England to France were grounded that day. But at the same time Miller left Twinwood Farm, Corrigan, then a captain, took off from RAF Station Fairford, eighty miles west-southwest of Bedford, flying a C-47 Dakota on a course that paralleled the Norseman's track. Corrigan agrees that the weather was terrible and that he flew "on the deck" until well into France, landing at Chateaudun, southwest of Paris. Edwards also learned that Colonel Corrigan had a conversation with a USAF major around 1953 relative to the Miller mystery. The major told Corrigan he had been stationed in France during December of 1944. He mentioned that, as a young military police lieutenant in Paris, he had been personally involved in the investigation of Glenn Miller's murder. The comment was not new to Corrigan; he recalled the stories told by aircrews returning from Paris leave who said rumor had it that Miller was found in Place Pigalle, Paris, with his skull crushed.

About this time, Edwards said, he received a photograph of Miller's aircraft—on the ground—at RAF Bovingdon, outside London. He was able to trace the eyewitness, he said, who also saw Glenn Miller boarding a Dakota at the same airfield that afternoon. It turned out that Miller was unhappy about flying in the small Norseman, Edwards said, because the discomfort from the rough air bothered him. The temperature changes during the cold weather conditions created turbulence because of the updrafts. It was such an uncomfortable flight, Edwards contended, that Miller changed his mind and decided he wanted to complete the flight by Dakota.

The three men took off from Twinwood Farm about one fifty-five in the afternoon and, by Edwards's calculations, they landed at RAF Bovingdon about two-thirty. Then the Norseman took off again about four P.M. to fly across the channel. But Miller was not aboard, Edwards asserted; he had crossed earlier by Dakota. The Edwards group traced the eyewitnesses who traveled with Miller on that flight and they concluded that he did, in fact, reach Paris.

According to Edwards, he and his associates managed to trace the [investigators in the] U.S. Military Police who allegedly investigated the murder of Miller in Paris and asked the U.S. Government for an official statement. Then, Edwards said, because he believed the evidence showed that Miller was murdered outside a red-light establishment in the Pigalle District on the fifteenth of December, 1944, he decided to kill the story in the press.

Investigators learned that the name of Major Alton Glenn Miller is inscribed on the World War II Tablets of the Missing, with the names of 5,125 other American servicemen, at the Cambridge American Cemetery, Cambridge, England. According to Colonel William E. Ryan, Jr., of the American Battle Monuments Commission, these are the names of those whose remains were not recovered or identified.

In May of 1979, Edwards said, a member of the Glenn Miller Society in the United States asked the Pentagon to disclose the government's information on the case under the public information law. He insisted that he be told what happened to Glenn Miller. The president's office, Edwards said, replied that they were not allowed to disclose how he died—at the request of Glenn Miller's estate.

Edwards was confident the Norseman crashed due to carburetor icing, as he and his associate had predicted, and believed the plane was under 150 feet of water. He said it lies eight and a half miles off Dunkerque and fourteen miles east of Dymchurch off the English coast. "By drawing a line from Dymchurch to Dunkerque and measuring those distances from either coast, one would have the exact position," he said. He assembled a team of divers to lift the wreck and a team of aircraft technicians to rebuild it when it was recovered from the Channel floor. But he was unable to find sponsors to provide the ten thousand pounds needed to bring it up. "I had to be honest with my potential sponsors," Edwards said, after he had reached the conclusion that Miller's body was not aboard the aircraft. "The project lost its significance and none of the sponsors were interested in doing it . . ." Edwards said at one time he intended to write a book about his investigation, but when he discovered what he believes actually happened to the bandleader he lost all interest. He decided it was time to end the quest and passed all of his material to a member of the Glenn Miller Society, including the names of those who investigated the alleged murder at the time, a picture of the Norseman just before it took off from Bovingdon, the sonar picture of the plane lying upside down on the sea bed, and the original weather reports marked *Secret*. Edwards concluded: "He also has the name of an American ex-WAAF who was in the control tower at the time Miller took off [from Bovingdon]. She now lives in Albuquerque."

A second conjecture centers on the possibility of an act inherent in all wars—mistaken identity. Was the Norseman shot down by an Allied fighter plane or by British anti-aircraft fire because it failed to give its IFF—identification friend or foe—radio signal as it flew outbound from the island? Was the flight, in fact, authorized? General Goodrich expressed shock when he learned that his personal pilot, with his subordinate officer's apparent concurrence, had flown his private, single-engine transport across the Channel.

Such an impulsive act by the venturesome Norman Baesell was not out of character. The colonel was described as a devil-may-care sort, a take-charge type. He liked music and a good time with the Miller band when they played at Milton Ernest Post. Baesell seemed to use the Norseman as often as did General Goodrich and, with the general confined to bed for several days, the airplane was at the executive officer's disposal.

In the early 1960s, British writer Ralph Barker, a member of the RAF during World War II, researched the Miller disappearance. He was puzzled by the depth of the mystery and, after much sorting of possibilities, he mentioned a suspicion—a probability—that the flight was never authorized. He said another persistent rumor kept surfacing that was particularly upsetting to the British. Still repeated in hushed confidence was the story from a member of the Eighth Air Force that Miller's Norseman was shot down by an Allied fighter plane, specifically a British fighter plane. Barker said his investigation showed, however, that according to records not a single air-to-air contact was reported on December fifteenth by any British units on the island: bomber, fighter or coastal commands, or by the 2nd Tactical Air Force. None had reported a contact or a claim.

But what of antiaircraft fire? While the poor weather might have discouraged the scrambling of fighter pilots for an air-to-air intercept, radar-directed antiaircraft guns could have tracked and hit a lone "unidentified" plane in the overcast. Or so could the antiaircraft weapons of an offshore naval vessel patrolling the channel. Is this what happened, and did the Army Air Force, realizing that a serious blunder had been made, supress the Intelligence report of the incident? Did the "Channel crash" account stem from the need of persons in high position to resolve the bandleader's disappearance quickly with the least international scandal? For some reason it was generally understood there was no court of inquiry held by RAF or USAAF intelligence. Did a court convene in secret, and were its findings withheld from publication? C. F. Alan Cass, archivist of the Glenn Miller Collection at the University of Colorado, says, however, that an official inquiry was held in Boston, in 1945, as to the findings of the major's death.

Certain other actions of the military—or lack of them—that immediately followed Miller's disappearance are still pondered today. Alan Stilwell, Miller's orderly, recalled: "In one to two days after he left, the M.P.s came to his room [and] took his clothes and belongings. His family never received them." Was the military withholding something? Here again Alan Cass disagrees. "The trunk that was taken up and sent away shortly after Miller's disappearance is today in New York City as part of the Glenn Miller estate," Cass said. "It has been inventoried—it really didn't have a lot of his personal effects—and described to me by people who are in charge of the estate."

Then there was the prophetic letter with an ominous overtone that Glenn sent to his brother in the States shortly before he departed for France. In part it said: *"By the time you receive this, we shall all be in Paris, barring a nose-dive into the English channel."* Tex Beneke, the band's best-known male vocalist, said in 1979: "Miller was scared . . . thought he would never get back." And Herb Miller, his brother, says today: "I feel we were fed a story . . . completely misled. [There] was no search . . . no officer in charge." But Royal D. Frey, curator of the Air Force Museum at Wright-Patterson AFB, says simply: "There was no time for a search; there was too much [war] activity."

The final possibility—a land crash in England—is founded in reports from persons who live in the area between Bedford and London. And also, notably, from a former WAAF aircraft controller who tracked the bandleader's flight from Twinwood. It comes as a highly credible account of Miller's final hours in England.

Late in 1978 the long-standing Miller mystery took on a new development—one that tied in closely with Twinwood Farm. A newspaper story told of Roger Barfoot, Secretary of the Chilterns Historical Society, who had located Royal Air Force records from RAF Halton, a base in Buckinghamshire. They indicated, Barfoot said, that a small aircraft had crashed in the Kimble area of the Chiltern Hills, near Princes Risborough, about the same time the Norseman disappeared. Barfoot believed this was the same plane that carried Miller and his two companions and suggested their remains may still lie in the heavy undergrowth of the English countryside—not in the English channel as generally presumed. Barfoot concluded the plane crashed into the ridge of low hills between Oxford and London while flying through the low overcast and the pilot was unable to send a distress signal. According to the news release Barfoot supported his belief with: "Miller's plane would have been in the area for the flight to Paris and there couldn't have been too many light planes around at that time and in those conditions."

The Chiltern Hills has been called "The Graveyard" by British pilots. During the war hundreds of Allied planes crashed there. Peter Halliday, leader of the Chilterns Aircraft Group, which locates and digs out the remains of these planes today, said thus far 120 crashed aircraft have been found in the Buckinghamshire area.

The case for a well-hidden wreckage of a small plane lying unnoticed these many years in the remote English countryside is not as far-fetched as it sounds. Wreckages of much larger planes of World War II vintage are even today coming to light in other lonely and little-traveled sections of the United Kingdom. Alexander Hamilton of Leighton Buzzard, Bedfordshire, author of *Wings of Night* and a former World War II pilot, recently told of a Lockheed Hudson twin-engine bomber that was discovered on June 24, 1979. It was pulled from a Scottish bog where it had crashed more than thirty-five years ago with its crew and disappeared.

The third possibility has aspects that are at variance with the departure of Glenn Miller as reported by Don Haynes. Clearly, there is another version as to what happened at Twinwood Farm that midday in December of 1944.

Mrs. M. L. "Dixie" Clerke lives today in British Columbia. During World War II, leading aircraftwoman Muriel Dixon of the Womens Auxiliary Air Force—the WAAF—was a control tower operator at Twinwood Farm. As a young Glenn Miller fan she was not only acquainted with his music, she had several opportunities to meet and talk with the musician through his band's use of the Twinwood airfield. About two o'clock on the afternoon of her twentieth birthday she was on duty when four American officers drove up in a staff car. She is clear and firm about what happened next for, telling about the experience in 1980, she said: "It made the day very unhappy for me."

From the balcony of the control tower she watched an argument between Major Miller and the other officers—Baesell and Morgan, and an unidentified junior officer whom she believes was a warrant officer. She suspects he might have been Paul Dudley, another member of the Miller organization. They stood beside the Norseman and apparently were having difficulty persuading the major to fly to Paris in the small utility plane. Miller wanted to fly in another, larger aircraft, a four-engine C-54 that was to be used by his band and which was parked on the perimeter of the field. Finally Miller agreed to go in the Norseman; the four officers climbed aboard, and the pilot started the engine. The WAAF controller gave the pilot taxi instructions and takeoff clearance. She is certain the weather at that time was good, not foggy as generally reported.

In the control tower Muriel Dixson monitored the plane's flight as it headed south toward the coast. She says that after about fifteen minutes the pilot radioed Twinwood to tell them they were going to transfer communications to ground control, and she next heard them talking to Bovingdon, outside London. After about another fifteen minutes she heard Bovingdon Ground Control give them permission to land there. "I know they landed there," she says, "for I heard the Norseman call from the perimeter [of the field] and ground control gave it permission to taxi.

"I was quite surprised about ten minutes later when I heard the same aircraft call for takeoff and receive clearance. I decided to keep listening until Bovingdon Control also gave them clearance out of their aircraft control circuit, after which I intended to turn my attention to other duties. It was about seven minutes later when I heard Bovingdon Control call the Norseman for a position report. There was no answer. They continued to call every three to four minutes for the next half hour."

At first Muriel Dixson thought the problem was a malfunction of the plane's radio, a situation not unusual in that day, and she knew the pilot would immediately land at the nearest base, in this case Bovingdon. It was dark by now and she thought too that perhaps they had been forced to land in a remote farm field and were awaiting daylight. She remained at the station when her duty time ended, listening for news of contact.

"I went off duty," she relates, "and returned at eleven

o'clock that evening. The first thing I said to the girl I was relieving was: 'Have they heard from Glenn Miller's aircraft?' She replied, 'No, they're still calling.' I telephoned several stations along the path the plane was to have flown and checked with radar posts and control towers all the way to the coastal watch, but no one had seen or heard the plane. I thought this was unusual, for a Norseman had an unusually loud and distinctive pitch to its engine and would have been heard even in an overcast. The next day I made further contacts with the stations but they still had nothing to report.''

Today Mrs. Clerke remains as convinced as she was thirty-seven years ago that the answer to Glenn Miller's disappearance lies in a time factor. "The plane didn't respond to Bovingdon's radio calls very soon after it left the field, and it didn't clear the circuit. In clear weather one could see to the edge of the control circuit from the tower. There was such a short period of time between its departure and Bovingdon's attempt to contact them that they must have crashed on land—not far from Bovingdon. The plane never reached the coast.

"Since that time I have learned about the reputation of the Chiltern Hills and realized if they did go down it had to have been there. One runway at Bovingdon lies in a direction that would take a plane right over the Chilterns. I'm told it was foggy around London that night and if the Norseman struck one of the cliffs the loose shale would have fallen down and probably buried it straightaway. I believe searchers will eventually find the aircraft there.''

The Clerke account is credible and profoundly significant. It could explain why none of the official announcements of Miller's disappearance say—or even suggest—the plane went down in the channel. It comes close to solving the Miller mystery, and would, if it could, also answer these questions: Why did the plane land at Bovingdon? Was it to drop off Paul Dudley, who survived the war? And was Miller himself, in fact, aboard the Norseman when it departed that station? Why would officials want to keep a land crash secret? Why, if Morgan, Baesell, and Miller were aboard the plane, would their place of burial also remain a secret?

Although Glenn Miller was no longer with his band, his musicians kept the promise he made to the GIs and broadcast live from Paris on Christmas Day. They continued to perform in the European theater until the war ended, then returned home and were disbanded.

Twelve years after Miller's disappearance the Glenn Miller estate authorized the re-formation of the Glenn Miller Orchestra under the direction of Ray McKinley. Since then it has had three directors of which Jimmy Henderson was the latest.

General Jimmy Doolittle, Commander of the Eighth Air Force, once told Glenn Miller in England that, next to a letter from home, his American band of the AEF was the greatest morale builder in the European theater. Miller was proud of his group's contribution to victory; they took a much-needed touch of the American homeland to weary GIs.

Speculation still abounds, and many believe that eventually the authorities somewhere—in America or in the United Kindom—will reveal the details of Glenn Miller's disappearance. Perhaps it will happen when all of those who have knowledge of the circumstances have at last passed on and are forever removed from the possibility of censure by the great bandleader's faithful and spirited following.

Norman Baesell's retort to Glenn Miller: *"Do you want to live forever?"* was ironic. No one realized more than Miller that he wouldn't live forever—but what he didn't know then was that his music would.

15

The Mystery of Flight 19

Throughout the history of manned flight, aircraft have disappeared without a trace. Pilots are reluctant to discuss these cases with outsiders, and will only guardedly talk about them among themselves.

They can understand crashes. As shocking as wreck sites are, the planes, or at least parts of them, are still in evidence to offer clues on the probable causes. But in the case of an airplane simply vanishing, a deep-seated, almost superstitious dread replaces the power of reason.

The disappearance of a squadron of torpedo bombers off the east coast of Florida remains one of the most incredible mysteries of our time. To confound the puzzle, a huge PBM-5 Marine flying boat sent to search for the missing planes vanished minutes after takeoff. Twenty-seven men were plucked from the skies under circumstances so unusual that, thirty-five years later, the sea has not given up a single clue as to their fate, nor has one fragment of wreckage been washed ashore or found afloat.

Soon after daybreak on Wednesday, December 5, 1945, a cloud cover moved out of the Gulf of Mexico and drifted across the Florida peninsula. It crossed the west coast between Tarpon Springs, south to Sarasota, and by midmorning the cloud bank reached the Atlantic shore between Daytona Beach and Melbourne. At one o'clock in the afternoon the overcast had all but dissipated over the open water east of Canaveral. Except for a few scattered clouds, most of the Florida peninsula was clear. Flight conditions

THE MYSTERY OF FLIGHT 19 283

improved rapidly. The warm Gulf Stream again became a bright blue and the sun lighted the sandy shorelines and lush, semitropical forests.

The stage was set for Flight 19 to make its never-to-be-forgotten journey to oblivion.

This unlikely story began that afternoon at the Fort Lauderdale Naval Air Station. Although the war had been over for three months, patrol flights were still conducted as part of training exercises in coastal defense. A squadron of five Grumman Avengers waited on the operations ramp as fourteen airmen strolled from the briefing room with navigation gear and flight charts and climbed into their cockpits. Each chart was marked with a triangle, one leg of which extended east to sea for 160 miles. At the end of this line, which would put the patrol in the vicinity of the Bahamas, they would follow the second line due north for forty miles, then swing southwestward toward the air station on the final course for home. It was to be a routine training exercise.

At two past two P.M., the flight leader closed his cockpit canopy, poured the power to the 1,750-horsepower Wright engine, and led the other four TBMs down the runway. This was the lead Avenger, FT-28, piloted by Navy Lieutenant Charles C. Taylor, with E-4 Walter R. Parpart as radioman and E-4 Robert F. Harmon as gunner. Marine Captain Edward J. Powers was in command of FT-36, with Marine Staff Sergeants H. O. Thompson and G. R. Paonessa as crewmen.

Torpedo bomber FT-117 was flown by Marine Captain George W. Stivers with crewmen Private R. P. Gruebeli and Sergeant R. F. Gallivan. Patrol plane FT-3 was piloted by Ensign Joseph Bassi. Seamen H. A. Thelander and B. E. Baluk, Jr. made up his crew. The remaining Avenger, FT-81, was flown by Marine First Lieutenant Forrest J. Gerber, with Marine Pfc. William E. Lightfoot as his only crewman. Normally a TBM carried a crew of three: pilot, gunner, and radio operator. But one crewman failed to report, and at eight past two, the five planes closed into a tight formation at

150 miles per hour over the sparkling blue Atlantic—one man short.

All the planes were mechanically sound: engines, controls, and navigations systems were in normal working order. Their tanks were full and all of the bombers carried self-inflating life rafts. Each crewman wore a life jacket and a parachute. All fourteen men had had flight experience ranging from thirteen months to six years. The weather was clear, and at sixty-five degrees, somewhat cool for southern Florida, even for December. A moderate to fresh breeze blew from the northwest, but the usual afternoon buildup of cumulus clouds, a characteristic of that semitropical climate, was absent.

One hour and thirty-five minutes later, at three forty-five P.M. to be exact, when the five-plane patrol should have been homeward bound on its final leg, the control tower at Lauderdale Naval Air Station heard a distress call.

"Flight 19 to Lauderdale Naval Air Station, this is an emergency," an uneasy voice radioed. "We appear to be off our course . . . cannot see land . . . repeat, cannot see land."

"What's your position?" the tower requested.

"Not certain of our position," the patrol leader replied. "We aren't sure . . . we seem to be lost . . ."

The tower operators were puzzled. How could five experienced pilot-navigators be lost at the same time, less than an hour's flying time from base in clear weather? The tower ordered the leader to perform the standard emergency procedure for all aircraft lost off Florida's east coast. "Fly due west," the operator instructed, knowing the planes must sooner or later cross the coastline.

The tower waited for an acknowledgment from Flight 19, and when none came, radioed, "Go ahead, Flight 19." There was a long pause before the squadron replied. The men in the tower were stunned by the flight leader's words.

"We don't know which direction is west. Everything is wrong . . . the sky looks strange. We aren't sure of any direction. Even the water doesn't look as it should!"

THE MYSTERY OF FLIGHT 19

Naval officers at the air station were unable to understand how an unusual appearance of the sky and water could cause the flight to become lost. Although a small magnetic storm in the local area might cause their magnetic compasses to go awry, the radio navigation gear in the planes would have given them a radio fix on Lauderdale Naval Air Station and, above all, the late afternoon sun was boldly visible on the western horizon. At Lauderdale it remained bright in the sky. What could be shielding the sun from the pilot's view two hundred miles at sea?

A hectic hour and forty-five minutes followed. While the tower operators listened to an exchange of confused chatter between the planes, they heard conversations that ranged from anxiety and near hysteria. Then, a few minutes after five P.M., flight leader Taylor did a strange thing; he turned command of the patrol over to senior Marine pilot Captain Stiver. At five twenty-five, fifteen minutes after the time scheduled for the squadron to be safely on the ground, the new patrol leader contacted the tower with a final, wavering call.

"We're still not certain of our position . . . have gas for seventy-five minutes more . . . can't tell whether we're over the Atlantic or the Gulf of Mexico . . . we think we must be about seventy-five miles northeast of Banana River and about 225 miles north of base. . . . It looks like we are . . ."

Abruptly, the message broke off. When the tower operators failed to reestablish contact, they alerted the Banana River air-sea rescue unit. Two search and rescue amphibians were manned by seven twenty-seven P.M. One of the huge gull-wing Martin PBM-5 Mariners, commanded by Lieutenant W. G. Jeffrey, nosed out of its berth with thirteen men, full rescue equipment and survival gear, and headed for the patrol's last assumed position. Once the twin-engine aircraft were airborne, Lauderdale tower radioed a message to the squadron that help was on the way. It was not acknowledged.

Lieutenant Jeffrey's twenty-eight-ton search plane had

exchanged several radio messages with Lauderdale NAS as it taxied out for takeoff, but twenty minutes en route, when it was presumed near Flight 19's probable position, the tower requested a report. There was no reply—then or ever. After an extended night sweep of the water off Florida, the other Mariner returned safely to base.

In the gathering dusk it became disturbingly apparent that something frightening was happening out there. Flight operations at Lauderdale NAS flashed an alert to the Miami Coast Guard Station. A rescue plane took off, retraced the PBM's path and reached the position where it had apparently disappeared. It returned after dark and reported search results negative.

Officially, the last message was received at five twenty-five P.M., but unofficially, another call was heard at the time the Avengers' five-hour fuel supply would have been exhausted. One bomber was heard trying vainly to call another at four past seven P.M. Its signal was so faint that little could be understood except the flickering call letters "FT . . . FT . . . FT . . ."

Through the long winter night the Navy and Coast Guard put ships to sea from ports along the east coast of Florida. Watch officers scanned the horizon for signal flares from life rafts, but at dawn no lights had been reported anywhere. In the early morning hours the Chief of Naval Advanced Training at Jacksonville Naval Air Station appointed a board of inquiry as the escort carrier USS *Solomons* nosed into the alert area with thirty planes. By now, twenty-one vessels were fanning out in an ever-widening search, and by midday three hundred airplanes formed systematic grid patterns between Bermuda and Florida. The Royal Air Force dispatched two search planes from Windsor Field at Nassau, as the methodical crisscross of air and sea intensified during the afternoon. As the hours passed without a clue, search officers had growing doubts about being in the proper sector. Six planes down and not a scrap of wreckage? Impossible! But nightfall came and nothing was sighted by the air-sea flotilla.

THE MYSTERY OF FLIGHT 19

All of the next day, Friday, the growing search spread outward to include a two-hundred-mile penetration of the Gulf of Mexico. Low-flying reconnaissance planes made sweeps from Key West to Jacksonville, while twelve large parties made an onshore search of three hundred miles of coastline between Miami Beach and St. Augustine. The Everglades were surveyed. After each high tide and offshore blow, the outlying coastal islands and beaches were scanned for clues. Every life raft, bit of flotsam, and remnant of clothing washed ashore was examined until it was ruled out.

Military commanders could not believe that six planes could suddenly vanish without a trace over a relatively small area of the Atlantic. Commander Howard S. Roberts, executive officer of the Fort Lauderdale NAS, reported that, although the flight was under the direction of an experienced combat navigator, his pilots could have been blown off course by high winds. Miami weather reports listed freak winds with occasional gusts of forty miles an hour and higher, as well as showers and thunderstorms over the Banana River sector, when the five-twenty-five call was made to Lauderdale NAS. But the patrol pilots had made no mention of adverse weather. Although the Avengers could have been blown off course, they were tough combat aircraft that would be unaffected by rough weather. Structural failure was ruled out. And even if the TBM was not the most buoyant combat plane in the Navy's inventory, Commander Roberts assured the press that, when ditched, the planes would remain afloat long enough for each life raft to be launched and inflated.

"Each man has been so well trained in emergency ditching at sea," he explained, "they shouldn't even get their feet wet."

Each TBM had a two-way radio. How could all five planes disappear in one swoop without one radio operator sending a last-minute SOS, or a hint of approaching disaster? Did all five go down en masse in a mid-air collision while flying wing to wing in "dead man" formation? Were they in a blinding storm?

What about the missing Mariner? It was last heard from twenty minutes after it departed Banana River—about seven-fifty P.M.—but the last "unofficial" call from one TBM was forty-six minutes earlier, at four past seven, when its weak and faltering "FT . . . FT . . ." call letters were picked up. Did the flying boat meet the same fate as the patrol? There was an unusually mild sea, making it feasible for the Mariner to land. It also carried an emergency radio transmitter that it never used.

Waterspouts were unlikely. Widespread debris would have littered the surface of the water.

Every theory failed to account for the weird sky conditions described by the pilots and their failure to orient themselves. One possibility after another was investigated and refuted.

On the night of the disappearance, the merchant steamer *Gaines Mills* was at sea off Florida's east coast. It radioed the Navy that its crew had seen an explosion in the air about seven-fifty P.M. and what appeared to be an airplane spin down into the sea off New Smyrna Beach. This coincided with the approximate time the TBMs' tanks would have run dry. Could this have been the Mariner? The planes that combed the area at daybreak were unable to answer the question; they reported no wreckage, no fragments, no oil slicks.

If the patrol had flown west, it would have crossed the Florida coast or the Keys. Flying east, the pilots would have sighted the Bahamas with its twenty-five-mile-long Grand Bahama Isle. To the northeast were Andros and Great Abaco Islands, land masses impossible to ignore. Had the patrol struck out north or south over open water, the Florida mainland would have been visible at times. Despite high winds, the squadron could not have flown in one direction for an hour and forty-five minutes—from the time they declared their state of confusion until their last radio call—without sighting one familiar landmark. The compasses of all five planes must have been erratic, for if the error was identical,

THE MYSTERY OF FLIGHT 19

all five would have flown in a straight line and crossed land somewhere. Conclusion: the patrol flew in circles between the Bahamas and Florida.

The search, which covered 380,000 miles over land and water, ended on December eleventh, but Navy ships and aircraft that traveled the area regularly were ordered to remain on indefinite alert. Months later the Navy delivered its final report. They had not found a single clue as to the fate of the twenty-seven men. Not a scrap of the six planes was recovered. The board of inquiry reluctantly admitted that its members "were not able to make even a good guess as to what happened."

Over the past two decades, scientific discoveries have thrown some light on the baffling reports connected with the squadron that vanished. The blacking out of the sun and the strange appearance of the sea described by the naval aviators is now believed to have more bearing on the strange disappearance than was originally believed. Some careful research brings to light past cases where sudden darkness occurred on bright sunshiny days and no eclipse was noted. For example, on September 24, 1950, a large area of the United States saw a weird purple-blue sun filtering through an eerie atmosphere. An official explanation was released to the uneasy populace. The unnatural appearance of the sun was attributed to widespread forest fires in Alberta, Canada. Smoke and haze had lifted to a very high altitude and altered the color of the sunlight. On September twenty-sixth, the high "smoke and haze" appeared over Britain and the sun was seen as a blue-green ball.

The smoke theory had one serious flaw. While the prevailing winds were moving the alleged clouds eastward across the United States, the haze was also moving westward across the state of Washington and obscuring the sun. No follow-up theory on how a wind could blow in opposite directions at the same time was forthcoming.

At ten o'clock on the morning of December 2, 1904, darkness fell on Memphis, Tennessee. The sun simply failed to shine. Fifteen minutes later the unearthly blanket of darkness lifted. A similar thing happened at Aitkin, Minnesota, in April of 1889.

On the afternoon of March 10, 1886, there was a ten-minute blackout which occurred as a thick cloud mass carried an area of darkness across the United States from west to east. The small circle of intense gloom—a semisolid body moving between the sun and earth—blocked out the light as it passed from coast to coast in little more than three hours.

What is the significance of sudden midday darkness? Astronomers tell us that space is not so empty as it was once thought to be. Our earth is gaining weight daily as billions of microscopic particles are swept into our atmosphere. Some experts have suggested that there are great masses of opaque gas and dust drifting in outer space and that our planet, hurtling through these cosmic clouds, causes a sharp decrease in sunlight over a limited area. If this condition is cosmic in origin, it could point to a rare and little-understood phenomenon that electronic experts call "reduced binding" or, more commonly, "a hole in the sky."

Magnetic disturbances from above have occurred before in erratic, wavelike pulses. Special instruments show these anomalies to be roughly circular, usually no more than a thousand feet in diameter, and extending upward an indefinite distance. A magnetic tornado of sorts.

The late Wilbur B. Smith of Ottawa, former electronic consultant for the Canadian government, was once in charge of investigating these deviations. He used an instrument of his own design to trace and plot the magnetic patterns. In 1950 and 1951, he produced evidence that correlated sudden localized areas of magnetic twisters with certain unexplained plane crashes. He also showed the binding to be more common in the southern latitudes. Whether they weave about or merely drop down and fade away is still unknown. "When we looked for some of them a few months later, we could find no trace," Smith reported. "Some planes, of course, would not be affected by the conditions, others might fly to pieces in a storm of turbulence."

Although the U.S. Navy believes that there is not enough evidence to support any such "unlikely theory" as an atmospheric aberration, its one-time classified operation, *Project Magnet*, plotted them on a worldwide scale. During the 1950s and 1960s, Super Constellations with highly sensitive magnetometers—instruments that can measure the earth's magnetic fields—discovered that erratic magnetic forces from space frequented the Caribbean-Key West area.

The fact that a substantial number of military aircraft and commercial flights regularly passed unscathed through this region has not lessened the haunting threat. The five Avengers and the Mariner were only six of several aircraft that were lost in this general area. The "Bermuda Triangle," as it is called, lies roughly within a line drawn from Bermuda to Puerto Rico to the Florida coast (through the Bahama chain) and back to Bermuda. What makes the disappearances so incredible is the simple fact that the area is by no means an isolated one. The Caribbean and Florida coasts are well populated. So are the offshore islands. Day and night a stream of air and sea traffic shuttles between the islands and the mainland, over relatively short distances.

In December 1948, another mysterious disappearance was recorded in the triangle. Captain Robert Linquist of Fort Meyers and copilot Ernest Hill of Miami were about to complete a thousand-mile flight from San Juan, Puerto Rico, to Miami. Their DC-3, a charter plane of Airborne Transport, Incorporated, carried thirty-two passengers, including two infants. Shortly after takeoff, stewardess Mary Burks served coffee and cookies to the passengers, all of whom were returning from Christmas holidays on the island. The mood was light. Early in the flight, Linquist reported by radio that the passengers were singing Christmas carols. Now, hours later, they neared the mainland with the tired passengers lulled to sleep by the monotonous drone of the engines and the half-light of the darkened cabin. At four-

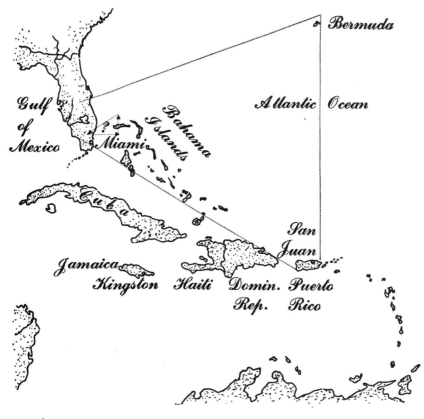

The Deadly Triangle. The small triangle shows the proposed flight path of the five Avengers of Flight 19.

THE MYSTERY OF FLIGHT 19

thirty A.M., Linquist contacted the Miami control tower. Ahead, dimly, the glow of the great city's lights edged the horizon.

"Approaching field. Fifty miles out. South. All is well; we'll stand by for landing instructions."

Seconds after Linquist made this call, it happened, and it was over quickly. There was no time for a distress call; the DC-3 had simply vanished.

What happened to the airliner, so near its destination? It was not forced down by bad weather; the Weather Bureau reported good flight conditions. Both pilots were experienced over the route, and there was no hint of mechanical trouble in flight.

There was a search, of course. Planes fanned into the area and scanned the water for debris and telltale packs of sharks and barracuda. Forty-eight vessels swept the region over seas so clear and shallow that anything as large as a transport plane could be seen on the bottom. After scouring 310,000 miles of the Everglades, the Keys, the Caribbean, and the Gulf of Mexico, the search was called off. To this day nothing has been found.

Twenty days later, on January 17, 1949, a British South American Airways plane vanished under equally baffling circumstances. It was the *Star Ariel,* an Avro Tudor IV en route from Bermuda to Santiago via Kingston. It carried thirteen passengers and a crew of seven commanded by Captain J. C. McPhee. The transport departed Bermuda at seven-thirty A.M., and climbed into a brilliantly clear morning sky. At eighty twenty-five, Bermuda tower heard this call: "This is Captain McPhee aboard the *Ariel* en route to Kingston, Jamaica, from Bermuda. We have reached crusing altitude. Weather fair. Expected time of arrival Kingston as scheduled."

This was the last call. Again, there was no wreckage, oil slick, clothing, or bodies. Nothing.

In the early 1950s, British Overseas Airways Company absorbed BSAA, and all records of the uncanny disappearance of the *Star Ariel* were turned over to the British Ministry of Aviation for thorough investigation. Their conclusions, not surprisingly, were almost identical with those of the U.S. Navy Board.

Five years later, in October of 1954, doom struck another aircraft, this time just north of the triangle. A Navy Constellation disappeared without sending an emergency call, despite having two powerful radio transmitters aboard. Several hundred planes and ships covered the area but, again, nothing was found. When questioned about the Constellation's disappearance, Commander Andrew Bright, director of the Navy's Aviation Safety Section, admitted officially there was "no explanation."

Two years passed before the invisible, lurking mantle of death visited the triangle again. On November 9, 1956, another Navy plane, a patrol bomber, disappeared off Bermuda. Again there was no radio call to warn of an impending disaster.

An Air Force KB-50 tanker left Langley Air Force Base, Virginia, on January 8, 1962, and headed for the Azores. The four-engine plane carried eight men under the command of Major Robert Tawney. A short time out of base, Langley tower heard a weak and wavering radio call from the tanker. Then the message drifted into silence. Contact could not be reestablished and the plane was presumed down at sea. Search units went into operation immediately, but, again, after seventeen hundred man-hours of combing the Atlantic, no trace of wreckage or bodies was found.

The next strike swallowed up two planes—KC-135 jet strato-tankers. On August 28, 1963, they took off in clear weather from Homestead AFB, Florida, to fly a classified refueling mission over the Atlantic. The crews totaled eleven men. At noon the planes radioed their position as eight hundred miles northeast of Miami and three hundred miles west of Bermuda. The KC-135s failed to make scheduled follow-up position reports and airway radio communications centers tried in vain to raise them. An air-sea search was

THE MYSTERY OF FLIGHT 19

ordered and the next day search planes sighted floating debris 260 miles southwest of Bermuda, but no survivors or bodies. A mid-air collision was first thought to have caused the crash, despite the statement from an Air Force spokesman who said the tankers were not flying in close formation and were in constant contact with one another. He was proved correct the following day when debris of the second tanker was found 160 miles from the first. No bodies were recovered.

Shortly before midnight on March 23, 1965, an air disaster again struck within the triangle. It happened during the joint Canadian-U.S. Exercise, *Maple Spring*.

One of three long-range RCAF sub hunters, a four-engine Argus, was operating sixty miles north of San Juan. It carried a crew of sixteen, which included three pilots, three navigators, two flight engineers, and seven electronic-equipment operators. One civilian, a scientific consultant to the Maritime Air Command, was also on board. The Argus, out of the U.S. Navy Base at Roosevelt Roads, San Juan, was on patrol four hours when the target submarine *Alcide* reported a flash on the horizon. She got under way immediately, along with Canadian Navy destroyers *Gatineau* and *Terra Nova*.

When the *Alcide* reached the scene, the crew recovered aircraft wreckage, rubber dinghies, and life jackets. An RCAF observer identified them as the type carried by the Argus.

A C-119 "Flying Boxcar," with ten men aboard, took off from Homestead Air Force Base at seven forty-seven P.M., the following June fifth to deliver a replacement engine to another C-119 stranded on Grand Turk Island. The 580-mile flight to the Bahamas was expected to take three and a half hours, and for Major Louis Giuntolli, the pilot, it was another routine mission. The flight path took them southeast along a heavily traveled air corridor called the "Yankee Route." The copilot, Lieutenant Lawrence Gares, settled the ungainly plane on the course plotted by navigator

Captain Richard Basset and made the usual radio position reports. At four past ten, they reported their position as over Crooked Island, about ninety-five miles northwest of their destination. This was about an hour before they were to touch down on Grand Turk. It was a routine call; no trouble was reported.

When the cargo plane was overdue at eleven twenty-three P.M., the Grand Turk control tower tried to raise them but failed. At five-thirty A.M., when the C-119's nine hours of fuel would have been exhausted, search units were already in action. Twenty-two Coast Guard, Navy, and Air Force planes swept the Bahama chain, while the cutters *Aurora* and *Diligence* plowed through seas that grew rougher by the minute. Squalls and low visibility impaired search efforts for several days. By the time the unsuccessful search was called off five days later, thirty-three planes were involved. On June tenth, the Coast Guard announced, "Five days and nights of searching. Results negative. There are no conjectures." The Air Force concurred with a statement that the ten men were "presumed dead."

A paradox was the report of another C-119 that passed over the same route in the opposite direction within an hour of Major Giuntolli's four-past-ten call. The crew agreed they encountered their best weather in the area where the C-119 was missing.

Although the waters of the deadly triangle are well patrolled by the Coast Guard, Navy, and Air Force, there is no reason to believe the weird disappearances of planes and ships will cease. Since 1966, an estimated four boats and two planes a month have disappeared in this graveyard of ships and aircraft. Many disappearances are lightninglike, sudden. The U.S. Seventh Coast Guard District, which is responsible for patrolling the area, is the busiest in the world, with ten thousand calls for assistance each year.

An impressive fleet of aircraft has vanished, without apparent cause, over this relatively small area of the Atlantic. Of the military and commercial planes, all were flown by

THE MYSTERY OF FLIGHT 19

experienced airmen and directed to their destinations by trained navigators. All carried radio and survival equipment and all disappeared in good weather. Almost all were daytime disappearances.

Until a survivor can tell the tale of what happens out there, the unpredictable whims of this strange force will remain a threat to seamen and airmen alike, not only within the Bermuda Triangle, but in other places of the world as well.

From July 20, 1927, until April 30, 1928, Charles Lindbergh made a series of long cross-country "good will" flights. They were sponsored by the Guggenheim Foundation for the progress of aviation. Several of them were in Central and South American countries, at the invitation of their governments.

On December 13, 1927, he left Bolling Field, Washington, D.C., to fly to Mexico City, Guatemala, British Honduras, Nicaragua, Costa Rica, Columbia, Caracas, San Juan, Haiti, and Havana. On the night of February 12 and 13, 1928, he took off from Havana for a nonstop flight to St. Louis, Missouri. He was flying the *Spirit of St. Louis,* the plane in which he had made his Atlantic solo crossing nine months earlier. The plane was equipped with an earth inductor compass, a directional compass that depends for its indications on the current generated by a small coil revolving in the earth's magnetic field. It is not a magnetic compass such as the type used in cars, boats, and small planes today—although his plane had a magnetic compass as well. Lindbergh's logbook entry for that flight through the darkness reads:

> Havana to Lambert Field, St. Louis, Missouri: 15 hours and 35 minutes. Both compasses malfunctioned over Florida Strait, at night. The earth-inductor needle wobbled back and forth. The liquid compass card rotated without stopping. Could recognize no stars through heavy haze. Located position, at daybreak, over Bahama Is-

lands, nearly 300 miles off course. Liquid compass card kept rotating until the *Spirit of St. Louis* reached the Florida coast.

Years later, during Lindbergh's days of semiretirement, the mysterious body of water off east Florida came into public notice. Did the Lone Eagle—an aerial navigator unmatched in his day and time—ever ponder over that night flight when he was three hundred miles off course, in the same area of the Bermuda Triangle that claimed Flight 19?

Index

Adams, Ensign, 47ff.
Akiyama, Josephine Blanco, 226f.
Allan, George "Scotty," 178, 182, 184, 186, 188f., 192
Aloha, 147, 149ff.
Amran, Biliman, 239
Andrée, Salomon (Andrée's polar expedition): burials, 114f.; crew, 91, 93; departure, 93f.; discovery, 100; flight, 103ff.; messages from, 92, 99; preparations, 88ff.; probable causes of deaths, 109ff.
Angel, Jimmie, 171f.
Angel of Los Angeles, 145

Baby, Raymond, Dr., 238
Baesell, Norman, Lt. Colonel, 262ff., 269, 278, 281
Ball, Albert, Captain, 10ff.
Barfoot, Roger, 277
Barker, Ralph, 275f.
Bell, George, 168ff.
Beary, Vice Admiral, 232
Bermuda Triangle disappearances, 291ff.
Berthaud, Lloyd, 126ff.
Boelcke, Oswald, 3ff., 7, 37ff.
Bogen, Eugene, 224
Bozon-Verduraz, Lieutenant, 14
Briand, Paul L. Jr., Colonel, 226, 228
Bridwell, Paul, Commander, 234f.
Brittain, M. L., Dr., 223, 224f.
buoys, message, 91, 99

Cass, C. F. Alan, 276
City of Oakland, 145
City of Peoria, 143
Clerke, M. L. "Dixie," 270, 278ff.
Cochrane, Jacqueline, 222f., 225f.

Cody, Lieutenant, 47ff.
Coli, Francois, 117, 119ff., 128
Connell, Dennis, 12f., 16f.
Corrigan, Tom, Lt. Colonel, 272
Cramer, Parker D., 135ff.
Crowe, Flight Leader, 10f.

Dallas Spirit, 147f., 152, 154f.
Dealy, Ted, 153f.
Dendrino, S., Lieutenant, 40
Dennison, Lee R., 169ff., 172
derelicts: Berserk Blimp, 47ff.; Boelcke's Derelict, 37ff.; Mindanao P-40, 40ff.
Desert of Thirst, 70
Deullin, Captain, 14ff.
Devine, Thomas E., 234
Dole, James Drummond, 142, 151, 155
Doyle, Sir Arthur Conan, 134f.
Dudley, Paul, Warrant Officer, 278ff.
Dunnell, Charles, 179

Eagle, 91ff.
Earhart, Amelia: *Amelia Earhart Lives*, 236f.; course error, 214; examinations of alleged remains, 235f., 238; Howland lighthouse, 239; line of position, 220; position report, Lae, 216; radio transmissions, 218ff.; request for bearing, 219f.; search for, 221ff.; theories as to fate, 229ff., 223
Earhart, Amy Otis, 223f.
Eckner, Hugo, Dr., 249, 258
Edwards, John, 270ff.
Endeavor, 132ff.
El Encanto, 146, 147

299

300 INDEX

Farrell, Bert, 180, 192
flares: Earhart search, 221; *Lady Be Good* site, 82; Mount Mauna Kea, 152; *Southern Cross Minor* site, 208f.
Fleming, Hans, Lt. Commander, 19
Flight 19: absence of survivors, 289, 293, 295, 297; crews, 283; distress calls, 284f.; flight route, 283; searches, 285ff.; theories on disappearances, 287ff.
Fokker, Anthony, 4, 6ff.
Forsmann, Swedish designer, 33ff.

Gambler's Chance, 129ff.
Garapan cemetery, 233ff.
Glasgow, Elsie May, 179, 192
Goerner, Fred, 233f., 235f.
Golden Eagle, 147, 149ff.
Griswolde, Tracy, Captain, 238
Guynemer, Charles, Captain, 14, 16ff., 122

Hailer, Carl, 12f.
Hamilton, Leslie, Captain, 124f.
Harrington, J. D., CPO, 230f.
Haynes, Don, Lieutenant, 260, 262ff., 278
Heinemann, Corporal, 8f.
Henderson, G. L. P., Colonel, 134
Hill, J. D., 126ff.
Hinchliffe, Walter, Captain, 132ff.
Hindenberg: crew and passenger escapes, 247f.; dimensions, 243f.; explosion and crash, 247f.; investigations and theories, 250ff., 256f.; landing approach, 246f., 251, 253; luxury, 243f.; safety record, 243f., 258; warnings, 250, 253
Hoeffel, Rear Admiral, 232
Hood, Charles Clyde Bronson, 180, 197
Hughes, F. L., Sergeant, 39, 40
Humming Bird, 144

Immelmann, Max, 4, 7ff.
Inglis, Donald, Lieutenant, 32f.

Jinx Flight: bail-out, 58f.; crew, 50f.; mechanical failures, 53ff.; mission, 51; reflections of, 66f.; reunions, 61ff.

Kanna, Ralph, 225

Kingsford, Smith, Charles, Sir: disappearance, 194; formation of ANA, 175; knighted, 194; long distance flights, 175, 193f.; search for *Southern Cloud,* 184ff.
Kinley, Robert, 225
Koelle, Captain, 20
Kothera, Don, 237

Lady Be Good: crew, 71f.; discovery of, 68ff.; finding of survivors, 80ff.; "jinx," 82f.; missing equipment, 73, 84; remains today, 83f.; searches for survivors, 72ff., 81f.; state of preservation, 69f., 72; unsolved mysteries, 84
Lady Southern Cross, 194
Lancaster, William Newton, Captain: and Chubbie Miller, 202; crash, 207; departure and flight, 203ff.; discovery of remains, 201f.; disposition of wreckage, 211f.; log book, 202, 208f.; search by French, 209
Lehmann, Ernst, Captain, 28, 245, 248ff.
Lindbergh, Charles, 123, 239, 297ff.
Loewenstein-Wertheim, Princess, 124f.
Loomis, Vincent, 238
Lowen, Fred, 168ff.
l'Oiseau Blanc (The White Bird), 117ff.

MacDonald, H. C., Lt. Commander, 129ff.
Mackay, Elsie, Honorable, 132ff.
mail route pioneers, 135ff.
Mannock, Edward Corringham, Major, 22ff.
Mantz, Paul, 232
Margules, Julian, 180
McClendon, Dennis E., Colonel, 84
McCown, Theodore, Dr., 235
McCubbin, G. R., Lieutenant, 8
McGoofma, Anna, Mrs., 237f.
Menckhoff, Karl, 17
Metcalf, J. D., 128f.
Mientjes, Flight Leader, 10
Miller, Glenn: band of AEF, 260ff.; departure for Paris, 264f.; disappearance, 267ff.; personality, 259f.; postwar band, 281; recordings, 259, 262, 268; rumors and theories, 269f., 275f.; Tablets of Missing, 274; unique musical sound, 259

INDEX

Minchin, Frederick, Colonel, 123f.
Miss Doran, 147ff., 155
Miss Hollydale, 145
Miss Southern Cross, 194
Mollison, Jimmie, 178, 181f., 187, 189
Morgan, John, Flight Officer, 264ff., 278, 281

Nimitz, Chester A., Admiral, 236
Norris, Curt, 43f.
Nungesser, Charles, 117ff., 128

Oklahoma, 146, 148
Old Glory, 126ff.
Oldmeadow, Rivers, Squadron Leader, 134
Operation Climax, 81
Operation Earhart, 226ff.
O'Reilly, Bill, 180

Pabco Pacific Flyer, 147f.
Pacquette, Oliver, 136ff.
Pellegreno, Ann, 236
phantom fighters of the blitz, 45ff.
Poll Bomber (Forsmann Giant): abandonment, 36; construction, 31ff.; performance, 31ff.; remains, 33; use as psychological warfare weapon, 27f., 36
Port of Brunswick, 159, 161, 163, 173
Project Magnet, 291
Pruss, Max, Captain, *Hindenberg* commander, 245ff., 255, 257f.

Redfern, Gertrude, 158, 161f., 165f., 173
Redfern, Paul: aircraft modifications, 159f.; declared forced down, 165; departure for Rio, 161ff.; early flying, 157f.; flight course, 161, 163ff.; searches and expeditions, 165ff.; sighted by *Christian Krogh*, 165
Reisenflugzeug, 26, 30. See also Poll Bomber
Richthofen, Lothar von, 10f., 12
Richthofen, Manfred von, 4, 37
Rosendahl, Charles, Commander, Lakehurst NAS, 245ff., 249, 251, 257

Sandy, J. L., Lieutenant, 39f.
Schmidt, Harry, General, 236

Schwonder, Captain, 20f.
Scott, Robert E., Brigadier General, 40ff.
Shortridge, William Travis, 178, 179, 189f., 191, 198
Sinclair, Gordon, Captain, 132
Sir John Carling, 128f.
Sonter, Tom, 195f.
Southern Cloud: accommodations, 176; "clues" and "finds," 195; crash site, 196ff.; crew, 178f.; disappearance, 183ff.; discovery, 195ff.; departure, 180f.; hoaxes, 192; influence on ANA, 175, 193; Memorial, 200; passengers, 179f.; psychic impressions, 181, 183, 186, 192, 196; reports of, 184, 186f., 189ff., 195, 198; searches abandoned, 191; searches made, 184ff.; specifications, 177; Sydney-Melbourne run, 181
Southern Cross, 175f., 194
Southern Cross Junior, 193
Southern Cross Minor, 194, 203, 207
Southern Moon, 176, 186, 188
Southern Sky, 176, 189
Southern Star, 176, 187
Southern Sun, 176, 186, 188, 190, 193
Spehl, Eric, 253, 255ff.
Spirit of Los Angeles, 144
Stokes, Claire, 179ff., 200
St. Raphael, 123ff.

Toole, Charles, Lt. Commander, 224
Tully, Terence, 128f.

Ulm, Charles, 175, 184, 186f., 194

Waller, J. H., Corporal, 8
Watson, John, 179, 181
Wissemann, Kurt, 17
Woolaroc, 145, 147, 150

Zaugg, D. J., Dr., 231f.
Zeppelins: London raids, 18, 28f.; L-44, 19; L-45, 19; L-49, 19; L-50, 17ff.; L-55, 19; L-71, 28f.; LZ-98, 245; LZ-127 *(Graf Zeppelin)*, 245, 250, 255, 257; LZ-129 *(Hindenberg)*, 243ff.; LZ-131 *(Graf Zeppelin II)*, 245, 257

About the Author

Dale Titler soloed thirty-eight years ago at Stultz Field, a small, grass airfield in Pennsylvania named in honor of Amelia Earhart's pilot on her 1928 trans-Atlantic flight. A flight instructor turned writer, Mr. Titler started early enough in flying to experience the waning days of aviation's open cockpit and biplane era—with its Travelaires, Wacos, OX-5 Challengers, and Sunday afternoon parachute jumps at the local airport.

During World War II he served with the Army Air Force and later graduated from the Pittsburgh Institute of Aeronautics. For ten years he taught aircraft engineering to Air Force officer pilots and aviation cadet trainees at Graham Air Base, Florida. From 1961 until 1978 he instructed Air Force students in aircraft crash firefighting and rescue operations, aircraft control and warning radar systems, and military personnel procedures. Mr. Titler is the historian for Keesler Technical Training Center. He lives with his wife in Gulfport, Mississippi.